T0207250

UNITEXT for Physics

More information about this series at http://www.springer.com/series/13351

Victor Ilisie

Concepts in Quantum Field Theory

A Practitioner's Toolkit

 Springer

Victor Ilisie
University of Valencia
Valencia
Spain

ISSN 2198-7882   ISSN 2198-7890   (electronic)
UNITEXT for Physics
ISBN 978-3-319-38723-9   ISBN 978-3-319-22966-9   (eBook)
DOI 10.1007/978-3-319-22966-9

Printed on acid-free paper

Springer International Publishing AG Switzerland is part of Springer Science+Business Media
(www.springer.com)

To my wife and daughter

Preface

This book is intended to be advanced undergraduate–graduate friendly. With a less strict yet formal language, it intends to clarify and structure in a very logical manner concepts that can be confusing in Quantum Field Theory. It does not replace a formal book on the subject. Its main goal is to be a helpful complementary tool for beginners and not-so-beginners in this field. The reader is expected to be at least familiar with basic notions of Quantum Field Theory as well as basics of Special Relativity. However, most of the times being familiar with Special Relativity doesn't mean being familiar with tensor algebra or tensor calculus in general. Many physics books assume that the reader is already familiar with tensors, so they begin directly with advanced topics. On the other hand, many mathematical books are somewhat too formal for a young physicist. Thus, I have introduced at the beginning, a nicely self-contained, student friendly chapter, which introduces the tensor formalism in general, as well as the concept of a manifold. This is done by assuming only that the the reader is familiar with the notions of vectors and vector spaces. Key aspects of Special Relativity are also covered.

The kinematics needed for the most common relativistic processes is given. It is a logical schematic list of all the relevant and most important formulae needed for calculating relativistic collisions and decays. It includes one-to-two and one-to-three body decays, and also the two-to-two scattering process both in the center of mass and laboratory frames. It also includes simplified general formulae of one, two, and three-body Lorentz invariant phase space. As a bonus, the three and four-body kinematics in terms of angular observables is also presented.

Noether's theorem is mostly treated in the literature in a somewhat heuristic manner by introducing many ad hoc concepts without too many technical details. I try to fix this problem by stating the most general (Lorentz invariant) form of the theorem and by applying it to a few simple, yet relevant, examples in Quantum Field Theory.

I also try to introduce a simple and robust treatment for dimensional regularization and consistently explain the renormalization procedure step-by-step in a transparent manner at all orders, using the QED Lagrangian, which is in my opinion the most suitable from an academical point of view. I dedicate thus, one chapter in explaining the Dyson summation algorithm and try to clarify all possible confusions that may arise. Various renormalization schemes are also presented.

Infrared divergences, as well as the ultraviolet ones are also extensively treated. I explicitly calculate a few infrared divergent Green functions and show an explicit example of cancellation of infrared divergences (step by step) using dimensional regularization. Other interesting topics are also discussed.

Possible issues and confusion for tadpole renormalization are commented and some illustrative simple examples are given in Chaps. 7 and 9, where we also treat the renormalization of the W sector of the Standard Model. With the tools given here one should find it straightforward to calculate and renormalize any N-point Green function at one-loop level. A very short example of a two-loop calculation is also given.

Valencia   Victor Ilisie
July 2015

Acknowledgments

Many thanks to Prof. A. Pich for sharing his clear vision with me, for having the patience to answer to all my questions and doubts, and for teaching me all kinds of subtleties in Quantum Field Theory and renormalization. (Also, I have borrowed many of his notations and conventions). Many thanks to Prof. J.A. de Azcárraga and Prof. J.N. Salas and for their wonderful classes on tensors, manifolds, Relativity, group theory, and many other advanced topics in physics that have inspired the first chapter of this book. Specially many thanks to Prof. J.A. de Azcárraga for many helpful comments on this manuscript. Also many thanks to S. Descotes for introducing me to the realm of angular observables. I would also like to thank my colleagues G. Torralba, J.S. Martínez, A. Crespo, and P. Bellido for our endless talks on physics, tensors, Relativity, and life in general. A lot of the merit is theirs. Last but not least, I would like to thank my family, that has always been so supportive and taught me the most important lesson of my life, to never give up on my dreams.

This work has been supported in part by the Spanish Government and ERDF funds from the EU Commission [Grants FPA2011-23778 and CSD2007-00042 (Consolider Project CPAN)] and by the Spanish Ministry MINECO through the FPI grant BES-2012-054676.

Contents

Chapter 1
Vectors, Tensors, Manifolds and Special Relativity

Abstract Assuming that the reader is familiar with the notion of vectors, within a few pages, with a few examples, the reader will get to be familiar with the generic picture of tensors. With the specific notions given in this chapter, the reader will be able to understand more advanced tensor courses with no further effort. The transition between tensor algebra and tensor calculus is done naturally with a very familiar example. The notion of manifold and a few basic key aspects on Special Relativity are also presented.

1.1 Tensor Algebra

Before we get to define the notion of a tensor, which will arise naturally, it is important to start from the very beginning and remember a few basic notions about vector spaces and linear maps (applications) defined over vector spaces. We shall assume that the reader is at least familiar with vectors and vector spaces, so we shall try not to get into unnecessary details. Let's, thus, start by considering a vector space V_n of finite dimension n defined over the set of real numbers[1] $\mathbb{R}$. Given an arbitrary basis $\{e_i\}_{i=1}^n$ we can write a vector $v \in V_n$ as

$$\boxed{v = v^i e_i}.$$
(1.1)

One is probably used to see a vector written in the following form

$$\mathbf{v} = \sum_{i=1}^n v^i \mathbf{e}_i.$$
(1.2)

Here we will suppress the bold vector symbol and adopt the standard Einstein summation convention for repeated indices, so what we get is the compact form (1.1). Let's consider an invertible change of basis given by the matrix A ($\det(A) \neq 0$). We can relate the new basis with the original one by

---

[1] In general it could be defined over $\mathbb{C}$, but here, we are not interested in this case. Once the reader is familiarized with the notions presented here, it is easy to further study the generalization to complex spaces.

© Springer International Publishing Switzerland 2016
V. Ilisie, *Concepts in Quantum Field Theory*,
UNITEXT for Physics, DOI 10.1007/978-3-319-22966-9_1

$$e'_j = A^i_j e_i, \tag{1.3}$$

where the upper index of A stands for the row and the lower one for the column (remember that, if not stated otherwise, summation is always performed over repeated indices). Equivalently one can write the inverse relation

$$e_j = (A^{-1})^i_j e'_i. \tag{1.4}$$

Because v is an invariant quantity, it is straightforward to obtain the transformation law for the vector components

$$v = v^j e_j = v^j (A^{-1})^i_j e'_i = v'^i e'_i. \tag{1.5}$$

Thus, under an invertible change of basis (1.3) the vector components transform as

$$v'^i = (A^{-1})^i_j v^j, \tag{1.6}$$

or equivalently $v^i = A^i_j v'^j$.

1.1.1 Dual Space

The dual space of V_n, denoted as V_n^* is defined as the space of all the linear maps (applications) from V_n to $\mathbb{R}$:

$$\beta : V_n \to \mathbb{R}$$
$$\beta : v \mapsto \beta(v), \tag{1.7}$$

with the following property

$$\beta(\lambda_1 u_1 + \lambda_2 u_2) = \lambda_1 \beta(u_1) + \lambda_2 \beta(u_2), \tag{1.8}$$

$\forall u_1, u_2 \in V_n$ and $\forall \lambda_1, \lambda_2 \in \mathbb{R}$. The space V_n^* is also a vector space of dimension n and its elements are usually called covectors. Using an arbitrary basis $\{\omega^i\}_{i=1}^n$ an element $\beta \in V_n^*$ can be written as

$$\beta = \beta_i \omega^i. \tag{1.9}$$

Therefore, given an element $v \in V_n$ the linear map $\beta(v) \in \mathbb{R}$ can be explicitly written as

$$\beta(v) = \beta_i \omega^i (v^j e_j) = \beta_i v^j \omega^i (e_j). \tag{1.10}$$

In general, the quantity $\omega^i(e_j)$ depends on the chosen bases $\{\omega^i\}$ and $\{e_i\}$. However, there is one basis of V_n^* called dual basis of V_n, that has the following simple property

$$\omega^i(e_j) = \delta^i_j, \tag{1.11}$$

where δ^i_j is the Kronecker-delta defined the usual way ($\delta^i_j = 0$ if $i \neq j$ and $\delta^i_j = 1$ if $i = j$). We shall work from now on using the dual basis and instead of writing its elements $\{\omega^j\}$ we shall write them $\{e^j\}$. Thus using this new notation (1.11) turns into

$$e^i(e_j) = \delta^i_j, \tag{1.12}$$

and so $\beta(v)$ takes the simple form

$$\boxed{\beta(v) = \beta_i \, e^i(v^j \, e_j) = \beta_i \, v^j \, e^i(e_j) = \beta_i \, v^j \, \delta^i_j = \beta_i \, v^i} \tag{1.13}$$

Let's now deduce how the elements $\{e^i\}$ must transform under a change of the basis $\{e_j\} \rightarrow \{e'_j\}$ in order to maintain the duality condition:

$$e^i(e_j) = \delta^i_j = e'^i(e'_j). \tag{1.14}$$

Let's suppose that the transformation $\{e^i\} \rightarrow \{e'^i\}$ is given by an invertible matrix B ($\det(B) \neq 0$):

$$e'^i = B^i_l e^l. \tag{1.15}$$

Inserting (1.3) and (1.15) into (1.14) we easily get

$$e'^i(e'_j) = B^i_l \, e^l(A^k_j \, e_k) = B^i_l \, A^k_j \, e^l(e_k) = B^i_l \, A^k_j \, \delta^l_k = \delta^i_j. \tag{1.16}$$

Therefore, we obtain the following relation between the matrices A and B

$$B^i_k A^k_j = \delta^i_j \implies BA = I \implies A = B^{-1}. \tag{1.17}$$

where I is the $n \times n$ identity matrix. In conclusion, the components of the covector and the basis of V_n^* obey the following transformation rules

$$\beta'_i = A^j_i \beta_j, \quad e'^i = (A^{-1})^i_j e^j. \tag{1.18}$$

1.1.2 Covariant and Contravariant Laws of Transformation

Summing up, given $v \in V_n$ a vector ($v = v^i e_i$) and $\beta \in V_n^*$ a covector ($\beta = \beta_i e^i$), we have the following law of transformation for $\{e_i\}$ and $\{\beta_i\}$

$$\boxed{e'_i = A_i^j e_j \,, \quad \beta'_i = A_i^j \beta_j} \,. \tag{1.19}$$

We shall call this, the **covariant** law of transformation. For $\{e^i\}$ and $\{v^i\}$ we have found

$$\boxed{e'^i = (A^{-1})^i_j e^j \,, \quad v'^i = (A^{-1})^i_j v^j} \,. \tag{1.20}$$

We shall call this, the **contravariant** law of transformation. This is the reason why we use upper and lower indices, to be able to make the difference between covariant and contravariant quantities. However, we have to be careful because *not every element with an upper or a lower index is a covariant or contravariant quantity*. We shall see an explicit example within a few sections.

1.1.3 Theorem

For a finite dimensional vector space V_n, the dual space of its dual space V_n^*, (called double dual space, denoted as V_n^{**}) is isomorphic to V_n.

  This is just general algebra and we shall not be concerned about giving the proof here. The important thing that we need to learn from this theorem is that there is a one-to-one correspondence between V_n and V_n^{**}, thus, in what we are concerned, we learn nothing new from V_n^{**}. As a consequence, we can safely identify the vector space V_n with its double dual V_n^{**}. Because of this, V_n can be viewed as the space of all linear maps from V_n^* to $\mathbb{R}$, $v : V_n^* \to \mathbb{R}$. Therefore, if one identifies V_n with the dual space of V_n^* then one can make the following definition

$$\boxed{e_j(e^i) \equiv e^i(e_j) = \delta_j^i} \,. \tag{1.21}$$

Given this definition one can also make another one that will turn out to be useful

$$\boxed{v(\beta) \equiv \beta(v) = \beta_i v^i} \,. \tag{1.22}$$

  After this short reminder, we are now in position to define a more general element of algebra that generalizes vectors, covectors and linear maps. We are talking of course, about tensors. First we will need to introduce the tensor product.

1.1.4 Tensor Product

Given two vector spaces V_n and V_m of finite dimensions n and m, the tensor product is a map of the form:

$$\otimes : V_n \times V_m \; \to \; V_n \otimes V_m$$
$$\otimes : (u, w) \; \mapsto \; u \otimes w, \tag{1.23}$$

with the following properties:

1. $(v_1 + v_2) \otimes w = v_1 \otimes w + v_2 \otimes w,$
2. $v \otimes (w_1 + w_2) = v \otimes w_1 + v \otimes w_2,$
3. $\lambda(v \otimes w) = (\lambda v) \otimes w = v \otimes (\lambda w),$
4. $v \otimes w \neq w \otimes v, \tag{1.24}$

$\forall v, v_1, v_2, \in V_n, \forall w, w_1, w_2, \in V_m$ and $\forall \lambda \in \mathbb{R}$. Note that the commutative property doesn't hold for the tensor product by definition.

The product $V_n \otimes V_m$ is a vector space of dimension $n \cdot m$ and its elements are called tensors. If $v = v^i \, e_i \in V_n$ and $w = w^j \, e_j \in V_m$ then $q \equiv v \otimes w$ can be written as:

$$q = v \otimes w = v^i \, w^j \, e_i \otimes e_j \equiv q^{ij} \, e_i \otimes e_j. \tag{1.25}$$

The tensor product can be defined over any finite sequence of vector spaces, dual spaces or both. We can define for example[2]:

$$\otimes : V_n \times \cdots \times V_n^* \times \cdots \times V_n \times \cdots \times V_n^*$$
$$\to \; V_n \otimes \cdots \otimes V_n^* \otimes \cdots \otimes V_n \otimes \cdots \otimes V_n^*, \tag{1.26}$$

etc.

1.1.5 What Do Tensors Do?

Tensors are multilinear maps that act on vector spaces and their duals.
For example

$$V_n \otimes V_n : V_n^* \times V_n^* \; \to \; \mathbb{R}$$
$$u \otimes v : (\alpha, \beta) \; \mapsto \; u \otimes w \, (\alpha, \beta). \tag{1.27}$$

---

[2] We shall only be concerned with identical copies of vector spaces and their duals, therefore all spaces considered from now on will be of dimension n.

The quantity $u \otimes w \, (\alpha, \beta)$ can be expressed using dual bases as

$$
\begin{aligned}
u \otimes w \, (\alpha, \beta) &= u(\alpha) \, v(\beta) \\
&= u^i \, e_i \, (\alpha_j \, e^j) \, v^k \, e_k \, (\beta_l \, e^l) \\
&= u^i \, \alpha_j \, e_i(e^j) \, v^k \, \beta_l \, e_k(e^l) \\
&= u^i \, \alpha_j \, \delta_i^j \, v^k \, \beta_l \, \delta_k^l \\
&= u^i \, \alpha_i \, v^k \, \beta_k
\end{aligned}
\tag{1.28}
$$

However, not all tensors defined over $V_n^* \times V_n^*$ are of the form $u \otimes v$. The general way of defining a tensor will be given in the following sections.

1.1.6 Rank Two Contravariant Tensor

A rank two contravariant tensor, or a $(2, 0)$ tensor is a linear map of the form:

$$
\begin{aligned}
t &: V_n^* \times V_n^* \to \mathbb{R} \\
t &: (\alpha, \beta) \mapsto t(\alpha, \beta),
\end{aligned}
\tag{1.29}
$$

with the following properties:

$$
\begin{aligned}
&1. \ t(\lambda_1 \alpha_1 + \lambda_2 \alpha_2, \beta) = \lambda_1 t(\alpha_1, \beta) + \lambda_2 t(\alpha_2, \beta) \\
&2. \ t(\alpha, \lambda_1 \beta_1 + \lambda_2 \beta_2) = \lambda_1 t(\alpha, \beta_1) + \lambda_2 t(\alpha, \beta_2),
\end{aligned}
\tag{1.30}
$$

$\forall \, \alpha, \beta, \alpha_1, \alpha_2, \beta_1, \beta_2, \in V_n^*$ and $\forall \, \lambda_1, \lambda_2 \in \mathbb{R}$. It is straightforward to deduce that the following property also holds

$$
t(\lambda_1 \alpha, \lambda_2 \beta) = \lambda_1 \lambda_2 \, t(\alpha, \beta),
\tag{1.31}
$$

$\forall \, \alpha, \beta \in V_n^*$ and $\forall \, \lambda_1, \lambda_2 \in \mathbb{R}$. Given $\alpha, \beta \in V_n^*$ we can write $t(\alpha, \beta)$ as

$$
t(\alpha, \beta) = t(\alpha_i e^i, \beta_j e^j) = \alpha_i \, \beta_j \, t(e^i, e^j) \equiv \alpha_i \, \beta_j \, t^{ij},
\tag{1.32}
$$

where we have defined $t(e^i, e^j) \equiv t^{ij}$ as the tensor components related to the given basis. Thus, we can express t using a basis and the tensor product as follows

$$
t = t^{ij} \, e_i \otimes e_j,
\tag{1.33}
$$

so that,

$$
\begin{aligned}
t(\alpha, \beta) &= t^{ij}\, e_i \otimes e_j \, (\alpha_k e^k, \beta_l e^l) \\
&= t^{ij}\, \alpha_k \beta_l \, e_i \otimes e_j (e^k, e^l) \\
&= t^{ij}\, \alpha_k \beta_l \, e_i (e^k) e_j (e^l) \\
&= t^{ij}\, \alpha_k \beta_l \, \delta_i^k \, \delta_j^l \\
&= t^{ij}\, \alpha_i \beta_j
\end{aligned}
\tag{1.34}
$$

Whenever t can be separated as $t^{ij} = v^i w^j$ with $u, w \in V_n$ (meaning that $t = v \otimes w$, as in the previous section) it is said that t is a separable tensor.

1.1.7 Rank Two Covariant Tensor

A rank two covariant tensor, or a $(0, 2)$ tensor, is a linear map of the form:

$$
\begin{aligned}
t &: V_n \times V_n \to \mathbb{R} \\
t &: (u, v) \mapsto t(u, v),
\end{aligned}
\tag{1.35}
$$

with the same properties 1, 2 as in the previous case. Thus, we can express t using a basis as follows

$$
t = t_{ij}\, e^i \otimes e^j \,.
\tag{1.36}
$$

Therefore, given $u, v \in V_n$

$$
\begin{aligned}
t(u, v) &= t_{ij}\, e^i \otimes e^j \, (u^k e_k, v^l e_l) \\
&= t_{ij}\, u^k v^l \, e^i \otimes e^j (e_k, e_l) \\
&= t_{ij}\, u^k v^l \, e^i (e_k) e^j (e_l) \\
&= t_{ij}\, u^k v^l \, \delta_k^i \, \delta_l^j \\
&= t_{ij}\, u^i v^j
\end{aligned}
\tag{1.37}
$$

1.1.8 (1, 1) Mixed Tensor

We have to be somewhat careful when we defining mixed tensors. For example, we can define a (1,1) mixed tensor in two different ways. The first one

$$t : V_n^* \times V_n \to \mathbb{R}$$
$$t : (\alpha, v) \mapsto t(\alpha, v), \tag{1.38}$$

with t first acting on V_n^* and afterwards on V_n. It must be written as

$$\boxed{t = t^i{}_j \, e_i \otimes e^j}. \tag{1.39}$$

The other way of defining a $(1,1)$ tensor is

$$t : V_n \times V_n^* \to \mathbb{R}$$
$$t : (v, \alpha) \mapsto t(v, \alpha). \tag{1.40}$$

In this case t must be written as

$$\boxed{t = t_i{}^j \, e^i \otimes e_j}. \tag{1.41}$$

In order to avoid this confusion one usually leaves blank spaces in between the tensor indices, as it is done here, to indicate the order in which the application acts. Let's take one last example. Consider the map

$$t : V_n \times V_n^* \times V_n^* \times V_n \to \mathbb{R}. \tag{1.42}$$

Obviously t must be written as

$$t = t_i{}^{jk}{}_l \, e^i \otimes e_j \otimes e_k \otimes e^l. \tag{1.43}$$

It must be noted that if, for practical calculations, this order does not count, one usually forgets about the blank spaces. This is pretty usual in many calculations in physics, for example you will probably find the previous tensor components written as t_{il}^{jk}.

1.1.9 Tensor Transformation Under a Change of Basis

Let's consider a $(2, 0)$ tensor. Under a change of basis of the form (1.3) we have the following

$$t = t^{kl} e_k \otimes e_l = t^{kl} e_i' \otimes e_j' \, (A^{-1})_k^i \, (A^{-1})_l^j = t'^{ij} e_i' \otimes e_j'. \tag{1.44}$$

Thus, the law of transformation of a rank two contravariant tensor is:

$$\boxed{t'^{ij} = t^{kl} (A^{-1})_k^i (A^{-1})_l^j} \tag{1.45}$$

Obviously, for a rank two covariant tensor the law of transformation is:

$$\boxed{t'_{ij} = t_{kl}\, A^k_i\, A^l_j}$$ (1.46)

and for a (1, 1) mixed tensor we have:

$$\boxed{\begin{aligned} t'^i_{\ j} &= t^k_{\ l}\, (A^{-1})^i_k\, A^l_j \\ t'_j{}^{\,i} &= t_l{}^{\,k}\, (A^{-1})^i_k\, A^l_j \end{aligned}}$$ (1.47)

The generalization to (r, s) tensors (r-times contravariant and s-times covariant) is straightforward. A $(0,0)$ tensor is called a scalar (remains invariant under a change of basis). A vector is a $(1, 0)$ tensor and a covector is $(0, 1)$ tensor.

1.1.10 Intrinsic Definition of a Tensor

The transformation laws (1.45), (1.46), (1.47) reflect the intrinsic definition of tensors. In order to demonstrate that some quantity is a tensor it is sufficient to show that it obeys the tensor laws of transformation.

1.1.11 Tensor Product Revised

The tensor product is a way of constructing tensors from other higher rank tensors, not only from vectors or covectors as we have done previously. It is obvious that, if t is a (r, s) tensor and b is a (m, n) tensor then $t \otimes b$ is a $(r + m, s + n)$ tensor. For example, if $t = t^i_{\ j}\, e_i \otimes e^j$ is a $(1, 1)$ tensor and $b = b^k_{\ l}\, e_k \otimes e^l$ is also a $(1, 1)$ tensor then, $q = t \otimes b$ is a $(2, 2)$ tensor and it is explicitly given by

$$q = t \otimes b = q^i_{\ j}{}^k_{\ l}\, e_i \otimes e^j \otimes e_k \otimes e^l = t^i_{\ j}\, b^k_{\ l}\, e_i \otimes e^j \otimes e_k \otimes e^l.$$ (1.48)

1.1.12 Kronecker Delta

It is easy to prove that the Kronecker delta is a rank two mixed tensor. It is also symmetric $\delta^j_i = \delta^i_j$. Being a mixed tensor, in principle we should be careful with the index order and leave blank spaces. However, we shall continue using the simplified notation $\delta^j_i \equiv \delta_i{}^j \equiv \delta^j{}_i$ because, for practically most of our calculations, this order does not really count.

1.1.13 Tensor Contraction

Given a rank (r, s) tensor, we can construct a rank $(r - 1, s - 1)$ rank tensor by contracting (summing over) any upper (contravariant) index with any lower (covariant) index. For example, given a rank $(2, 2)$ tensor with components $T_{ij}{}^{lk}$, the quantities $T'^{k}_{j} \equiv T_{ij}{}^{ik}$ are the components of a rank $(1, 1)$ tensor.

Similarly, one can construct tensors by contracting any upper index of a tensor with any lower index of another tensor. Given a rank $(1, 2)$ tensor with components $T^{i}{}_{jl}$ and a rank $(3, 0)$ tensor with components K^{kmn}, the quantities $P^{i}{}_{j}{}^{mn} \equiv T^{i}{}_{jl} K^{lmn}$ are the components of a $(3, 1)$ tensor.

1.1.14 Metric Tensor

Given a vector space V_n, we define a metric tensor (rank two covariant) as:

$$g : V_n \times V_n \rightarrow \mathbb{R}$$
$$g : (u, v) \mapsto g(u, v) \tag{1.49}$$

with the following properties:

$$1.\,\text{symmetric} \quad g_{ij} = g_{ji}$$
$$2.\,\text{non-singular} \quad \det(g) \neq 0 \tag{1.50}$$

Of course, we can define the inverse metric tensor (rank two contravariant)

$$g^{-1} : V_n^* \times V_n^* \rightarrow \mathbb{R}$$
$$g^{-1} : (m, n) \mapsto g^{-1}(m, n) \tag{1.51}$$

with the same two properties as the metric tensor. Thus, $g\,g^{-1} = I$, which can be written using the Kronecker delta as

$$\boxed{g_{ij}\,g^{jk} = \delta^k_i} . \tag{1.52}$$

Again, I stands for $n \times n$ the identity matrix.

1.1.15 Lowering and Raising Indices

There is a natural way of going from a vector space to its dual by using the metric tensor. Given a vector $v = v^j\,e_j$ we can define the covector $v^* = v_j\,e^j$ with

$v_j = g_{ij}v^i$, and given a covector $\beta = \beta_i\, e^i$ we can define a vector $\beta^* = \beta^i\, e_i$ with $\beta^i = g^{ij}\beta_j$. This can be generalized for any (r, s) type tensor, and we can use the metric tensor to lower or raise as many tensor indices as we want, for example

$$R_{ij}{}^k = R^{mnk}\, g_{mi}\, g_{nj}. \tag{1.53}$$

1.1.16 Scalar Product

Given a metric tensor we can define two important invariant quantities (scalars). First, given two vectors u, v we define their scalar product as:

$$\boxed{(u \cdot v) = (v \cdot u) \equiv v^i\, u_i = g_{ij}\, u^i\, v^j = g^{ij}\, u_i\, v_j = \delta^i_j\, u_i\, v^j}. \tag{1.54}$$

Second, when $u = v$ can define the squared modulus of the vector $v \in V_n$ as

$$\boxed{v^2 \equiv v^i\, v_i = g_{ij}\, v^i\, v^j = g^{ij}\, v_i\, v_j = \delta^i_j\, v_i\, v^j}. \tag{1.55}$$

1.1.17 Euclidean Metric

Even though we do not write it down explicitly when working in the usual Euclidean 3D space $\mathbb{R}^3$, we use the Euclidean metric given by

$$g = g^{-1} = \begin{bmatrix} 1 & 0 & 0 \\ 0 & 1 & 0 \\ 0 & 0 & 1 \end{bmatrix} \tag{1.56}$$

(for the canonical basis and Cartesian coordinates). The short-hand notation for this is $g_{ij} = g^{ij} = \mathrm{diag}\{1, 1, 1\}$.

Invariant Euclidean Length: given two *points in space* A and B with *coordinates* given by $\mathbf{x}$ and $\mathbf{y}$ in a certain reference frame $\mathcal{O}$, we define $\mathbf{w} \equiv \mathbf{x} - \mathbf{y}$. The squared distance between these to points in 3D Euclidean space is defined as

$$w^2 \equiv |\mathbf{w}|^2 = (w^1)^2 + (w^2)^2 + (w^3)^2 = g_{ij}\, w^i\, w^j. \tag{1.57}$$

We can observe that the length we have just defined is basis independent (obviously this has to hold because the distance between two objects doesn't depend on the reference system, at least from a classical point of view).

As we have already mentioned before, **not everything that has an index is a tensor**. For example, the **position vectors x and y that we are so used to call vectors must strictly be called coordinates, because they do not behave as vectors**. If we make a translation from $\mathcal{O}$ to another reference frame $\mathcal{O}'$ so that:

$$x'^i = x^i + a^i, \quad y'^i = y^i + a^i, \tag{1.58}$$

it is clear that

$$|\mathbf{x}'|^2 \equiv (x'^1)^2 + (x'^2)^2 + (x'^3)^2 \neq |\mathbf{x}|^2 \equiv (x^1)^2 + (x^2)^2 + (x^3)^2, \tag{1.59}$$

and same for $|\mathbf{y}'|^2$. A properly defined vector is $\mathbf{w}$; under the transformation (1.58), $w'^i = w^i$ so $|\mathbf{w}'|^2 = |\mathbf{w}|^2$. This simple example (that can be easily generalized to any n-dimensional Euclidean space with any metric) will turn out to be very useful for the transition from tensor algebra to tensor calculus.

1.1.18 V_n, E_n *and* $\mathbb{R}^n$

In the previous example we have mentioned *points in space*, without giving any proper explanation. We shall not give it yet. Within a few sections we shall see that these *points* are related to the concept of *manifold*. Let us define E_n as *the n-dimensional Euclidean space (or manifold)* as the abstract set formed by the *points in space A, B...*, with coordinates given by sub-sets of $\mathbb{R}^n$. Even if there is a global one-to-one relation between the coordinates and the points (between E_n and $\mathbb{R}^n$), **we must not identify the points of the space (nor the space) with the coordinates**. Therefore, we will say that

$$\boxed{w \in V_n \, ; \quad x, y \in \mathbb{R}^n \, ; \quad A, B \in E_n} \, . \tag{1.60}$$

We shall see in a few sections why it is so important (crucial) to define these three **different** spaces.

1.2 Tensor Calculus

1.2.1 Tensor Fields

Whenever a vector is defined for every point in space $A \in E_n$, thus when it is a continuous function of some parameters $x^i \in \mathbb{R}^n$, it is called a vector field. How does it transform? (Obviously now the transformation matrix depends on the parameters x^i). In order to find the natural answer to this question let us take the following example. Consider again two points A, $B \in E_n$ with coordinates x^i, y^i in some reference frame and x'^i, y'^i in another one (related to the original one by a translation for example) (with x^i, x^i, x'^i, $x'^i \in \mathbb{R}^n$). Now let's define $\Delta x^i \equiv x^i - y^i$ and $\Delta x'^i \equiv x'^i - y'^i$. We have seen that the following interval is invariant:

$$\Delta s^2 = g_{ij}\,\Delta x^i\,\Delta x^j = \Delta s'^2 = g'_{ij}\,\Delta x'^i\,\Delta x'^j. \tag{1.61}$$

Taking x^i and y^i to be infinitesimally close (thus x'^i and y'^i) our scalar interval *becomes* differential

$$\boxed{ds^2 = g_{ij}\,dx^i\,dx^j = ds'^2 = g'_{ij}\,dx'^i\,dx'^j}. \tag{1.62}$$

From the previous expression it is natural to identify dx^i with the components of a vector field. So, under a change of coordinates $x^i \to x'^i$, the transformation law for the components of the vector field is given by the chain rule

$$dx'^i = \frac{\partial x'^i}{\partial x^j}\,dx^j, \tag{1.63}$$

which we identify as our contravariant law of transformation. This can be generalized to any tensor. If a tensor is a continuous function of some parameters $x^i \in \mathbb{R}^n$, then it is called a tensor field. Taking the usual two examples, $(2, 0)$ and $(0, 2)$ tensor fields can be written in the following form[3]:

$$t = t^{ij}(x)\,e_i(x) \otimes e_j(x),$$
$$l = l_{ij}(x)\,e^i(x) \otimes e^j(x). \tag{1.64}$$

Note that, necessarily the bases also depend on the same parameters x^i: for a point A with coordinates x^i_A in a reference frame, and x'^i_A in another, a tensor field evaluated in A denoted as t_A, must have a unique value which is independent of the reference frame. For example for a $(2,0)$ tensor field we have

$$t_A = t^{ij}(x_A)\,e_i(x_A) \otimes e_j(x_A) = t'^{ij}(x'_A)\,e'_i(x'_A) \otimes e'_j(x'_A). \tag{1.65}$$

A very familiar example where the components e_i of the basis depend on the coordinates x^j are vectors expressed in *curvilinear coordinates*. The basis is given by

$$e_1(x) \equiv \hat{u}_\phi, \quad e_2(x) \equiv \hat{u}_\theta, \quad e_3(x) \equiv \hat{u}_r. \tag{1.66}$$

with $x^j = (r, \theta, \phi)$.

In conclusion, taking quick look at (1.62) and (1.63), we identify the **contravariant** law of transformation with

---

[3]Here we will use the short-hand notation $f(x^i) \equiv f(x)$.

$$
\boxed{
\begin{aligned}
e'^{i}(x') &= \frac{\partial x'^{i}}{\partial x^{l}}\, e^{l}(x) \\
t'^{ij}(x') &= \frac{\partial x'^{i}}{\partial x^{l}}\frac{\partial x'^{j}}{\partial x^{k}}\, t^{lk}(x) \\
&\quad\cdots
\end{aligned}
}
\tag{1.67}
$$

Thus, the **covariant** law of transformation will be given by

$$
\boxed{
\begin{aligned}
e'_{i}(x') &= \frac{\partial x^{l}}{\partial x'^{i}}\, e_{l}(x) \\
t'_{ij}(x') &= \frac{\partial x^{l}}{\partial x'^{i}}\frac{\partial x^{k}}{\partial x'^{j}}\, t_{lk}(x) \\
&\quad\cdots
\end{aligned}
}
\tag{1.68}
$$

The law of transformation for mixed tensor fields is obviously given by

$$
\boxed{
\begin{aligned}
t'^{i}{}_{j}(x) &= \frac{\partial x'^{i}}{\partial x^{l}}\frac{\partial x^{k}}{\partial x'^{j}}\, t^{l}{}_{k}(x) \\
t'_{j}{}^{i}(x) &= \frac{\partial x'^{i}}{\partial x^{l}}\frac{\partial x^{k}}{\partial x'^{j}}\, t_{k}{}^{l}(x) \\
&\quad\cdots
\end{aligned}
}
\tag{1.69}
$$

The generalization to (r, s) tensor fields is straightforward. A $(0, 0)$ field is called a scalar field and it obeys:

$$
\boxed{\phi(x) = \phi'(x')}.
\tag{1.70}
$$

The expressions (1.67–1.70) represent the *intrinsic definition of tensor fields* and this is what you would normally find in many physics books. However, I believe that, in order to obtain a complete vision and a deeper understanding of tensors, one has to go through all the previous steps.

1.2.2 Tensor Density

Using the the intrinsic definition of tensor fields it is straightforward to introduce another object which is called *tensor density*. We say that $t^{i_1 \ldots i_r}{}_{j_1 \ldots j_s}(x)$ are the components of a (r, s) tensor density of *weight* W if under a change of coordinates $x^{i} \rightarrow x'^{i}$ they obey the following transformation law

$$
t'^{i_1 \ldots i_r}{}_{j_1 \ldots j_s}(x') = \left[\det\left(\frac{\partial x^i}{\partial x'^j}\right) \right]^W \frac{\partial x'^{i_1}}{\partial x^{l_1}} \cdots \frac{\partial x'^{i_r}}{\partial x^{l_r}}
$$

$$
\times \frac{\partial x^{m_1}}{\partial x'^{j_1}} \cdots \frac{\partial x^{m_s}}{\partial x'^{j_s}} \, t^{l_1 \ldots l_r}{}_{m_1 \ldots m_s}(x) \qquad (1.71)
$$

where $\det(\partial x^i / \partial x'^j)$ is the determinant of the Jacobian matrix of the given transformation. Same definition is valid for any (r, s) type tensor i.e., $t^{i_1 \ldots i_l \quad i_{l+1} \ldots i_r}{}_{j_1 \ldots j_k \quad j_{k+1} \ldots j_s}(x)$, etc. One can prove that the totally antisymmetric Levi-Civita symbol

$$
\epsilon_{i_1 \ldots i_k \ldots i_m \ldots i_n} = (-1)^p \, \epsilon_{i_1 \ldots i_m \ldots i_k \ldots i_n}, \qquad (1.72)
$$

where p stands for the parity of the permutation ($p = 1$ for an odd and $p = 2$ for an even permutation) is a tensor density of weight $W = -1$.

We will now move on to the next section and generalize everything to non-Euclidean spaces and introduce properly the concept of **manifold**.

1.3 Manifolds

We have defined the Euclidean space (manifold) E_n as the set of points that have global a one-to-one correspondence with $\mathbb{R}^n$. What is, however, the generic definition of a manifold and what happens if the manifold is not Euclidean? Hobson's wonderfully intuitive definition[4] is:

In general, a manifold is any set that can be continuously parametrized. The number of independent parameters required to specify any point in the set uniquely is the dimension of the manifold. [...] In its most primitive form a general manifold is simply an amorphous collection of points. Most manifolds used in physics, however, are "differential manifolds", which are continuous and differentiable in the following way. A manifold is continuous if, in the neighborhood of any point P, there are other points whose coordinates differ infinitesimally from those of P. A manifold is differentiable if it is possible to define a scalar field at each point of the manifold that can be differentiated anywhere. [...] An N-dimensional manifold $\mathcal{M}$ of points is one for which N independent real coordinates $\{x^i\}_{i=1}^N$ are required to specify any point completely.

For a generic manifold, one can only find small (local) mappings between the manifold and $\mathbb{R}^n$. A very familiar example is a surface; in order do describe a surface one needs two parameters, therefore we say that a surface is a two-dimensional manifold. Let's now imagine that our surface is flat. If this is the case we can find a global one-to-one mapping between the surface and $\mathbb{R}^2$. However, if we consider a sphere, in general one can never find a global mapping between this surface and

---

[4] M.P. Hobson, G.P. Efstathiou and A.N. Lanseby, General Relativity, *An Introduction for Physicists.*.

$\mathbb{R}^2$ that is able to cover the whole sphere. We can only locally describe parts of the surface using local *charts* (regions) of $\mathbb{R}^2$. This example can easily be generalized to any manifold.

An **N-dimensional differential manifold** $\mathcal{M}$ *is a set of elements (points) P, together with a collection of subsets* $\{\mathcal{O}_\alpha\}$ *of* $\mathcal{M}$ *that satisfy the following three conditions:*

1. $\forall\, P \in \mathcal{M}$ there is at least a subset $\mathcal{O}_\alpha$, so that $P \in \mathcal{O}_\alpha$. Equivalently:

$$\boxed{\mathcal{M} = \bigcup_\alpha \mathcal{O}_\alpha}$$

2. For each $\mathcal{O}_\alpha$ there is a diffeomorphism (differential with its inverse also differential) Ψ_α:

$$\boxed{\Psi_\alpha : \mathcal{O}_\alpha \;\rightarrow\; \Psi_\alpha(\mathcal{O}_\alpha) \equiv \mathcal{U}_\alpha \subseteq \mathbb{R}^n}$$

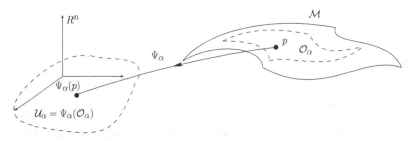

3. If $\mathcal{O}_\alpha \cap \mathcal{O}_\beta \neq \{\varnothing\}$ then $\Psi_\alpha(\mathcal{O}_\alpha \cap \mathcal{O}_\beta)$ and $\Psi_\beta(\mathcal{O}_\alpha \cap \mathcal{O}_\beta)$ are open sets of $\mathbb{R}^n$ and the application:

$$\boxed{\Psi_\beta \circ \Psi_\alpha^{-1} : \Psi_\alpha(\mathcal{O}_\alpha \cap \mathcal{O}_\beta) \;\rightarrow\; \Psi_\beta(\mathcal{O}_\alpha \cap \mathcal{O}_\beta)}$$

is a diffeomorphism. Here we have defined the composed operator $\Psi_\beta \circ \Psi_\alpha^{-1}(X) \equiv \Psi_\beta(\Psi_\alpha^{-1}(X))$. This application is called a *change of coordinates*.

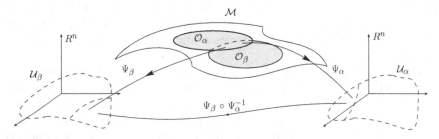

The sets $(\Psi_\alpha,\, \mathcal{O}_\alpha)$ are called *charts*. The set formed of all charts is called an *atlas*.

Thus, for a given manifold $\mathcal{M}$ of dimension n, if we can find an atlas that contains only one chart, then we shall say that the manifold is *flat* or that it has *trivial topology*.

Going back to the surface example, a surface has two types of curvatures, an intrinsic (Gauss curvature) and an extrinsic one. A surface that has no intrinsic curvature can be unfolded into a flat surface (for example a cylinder). On the other hand, a sphere is impossible to unfold into a flat surface (therefore we say it has an intrinsic curvature, or its topology is non trivial).

1.3.1 Embedding

Following these definitions, a curve is a one-dimensional manifold, a surface is a two-dimensional one and a solid volume is a three-dimensional one. All these examples are very familiar. Let's take for example the curve. We need one parameter to describe it, say t. We can describe this curve in $\mathbb{R}^3$ as:

$$\mathbf{r}(t) \equiv (x(t), y(t), z(t)). \tag{1.73}$$

If we have a surface, as we have mentioned before, we need two parameters to describe it, say u and v. In $\mathbb{R}^3$ we can describe a surface as:

$$s(u, v) \equiv (x(u, v), y(u, v), z(u, v)), \tag{1.74}$$

for example, we can describe regions of the given surface as $s(u, v) \equiv (u, v, f(u, v))$.

In the previous examples we have described (and visualised) these objects in $\mathbb{R}^3$ space. The technical word for *describing* (or *visualising*) these objects in $\mathbb{R}^3$ is **embedding**. We say that we can make and embedding of curves, surfaces or volumes in $\mathbb{R}^3$. However, not all one, two and three-dimensional manifolds are curves surfaces or solid volumes as we know them. There are for example, two dimensional manifolds that can not be embedded in $\mathbb{R}^3$ but in a higher dimensional space $\mathbb{R}^n$ with $n > 3$. Thus, these objects are not usual surfaces.

What happens if the manifold we are trying to describe has more than three dimensions? In General Relativity our manifold is formed by space-time points that we call *events*, it is four-dimensional and in general its topology is not trivial. As we have no information about the existence of any higher dimensional space in which our space-time can be embedded, our only solution is to abandon the embedding concept and describe our manifold intrinsically. In order to understand this we shall discuss the classical example. Let's imagine a civilization that lives in a two dimensional world which is a surface. The habitants of this surface can only measure the intrinsic curvature. Information about the extrinsic curvature is only accessible to an observer that lives in all tree dimensions. Similarly, we are the civilization that is living in four-dimensional space-time (which is our *hyper-surface*) and we have no access to a higher dimensional space (we don't even know if it exists). We can only make intrinsic measurements of the curvature of our manifold, which is related to the gravitational force.

1.3.2 Tangent Space $T_P(\mathcal{M})$

Consider the velocity vector of a moving body. This vector is tangent to the curve
that describes the trajectory of the moving body for every regular point of the curve.
A vector field defined over a surface is tangent to the surface for every regular point
of the surface. This vector field belongs to the *tangent plane* of the surface. What
happens for a general n-dimensional manifold? By analogy, we say that vector fields
belong to the *tangent space* of the manifold $\mathcal{M}$. The tangent space is defined for
every regular point of the manifold, $P \in \mathcal{M}$ and we will denote it by $T_P(\mathcal{M})$. We
shall call the union of all these spaces $T(\mathcal{M})$, thus:

$$v = v^i(x)\, e_i(x) \in T(\mathcal{M}); \quad v_P = v^i(x_P)\, e_i(x_P) \in T_P(\mathcal{M}). \qquad (1.75)$$

Thus, as usual we shall call v a vector field and v_P simply, a vector. Now we can
clearly see why we have insisted in making the difference between V_n, E_n and $\mathbb{R}^n$.
After extending our analysis to generic manifolds we can see it clearly:

$$\boxed{P \in \mathcal{M},\ x^i \in \mathbb{R}^n \text{ and } v \in T(\mathcal{M})}. \qquad (1.76)$$

These three spaces are, no doubt, different.

   As we have already seen, under a change of coordinates $x^i \rightarrow x'^i$, the bases of
the vector fields obey the covariant law:

$$e'_i(x') = \frac{\partial x^l}{\partial x'^i}\, e_l(x). \qquad (1.77)$$

In differential geometry it is usual to define this basis as the one formed by partial
derivatives that obey this exact transformation (1.77). Using this basis, a vector (field)
can be written as

$$\boxed{v_P = v^i(x) \left.\frac{\partial}{\partial x^i}\right|_P \in T_P(\mathcal{M}); \quad v = v^i(x) \frac{\partial}{\partial x^i} \in T(\mathcal{M})}. \qquad (1.78)$$

Thus we can interpret a vector v as map of the form:

$$\begin{aligned} v : \mathcal{F}(\mathcal{M}) &\rightarrow \mathbb{R} \\ v : f &\mapsto v(f), \end{aligned} \qquad (1.79)$$

with $\mathcal{F}(\mathcal{M})$ being the space of the all differentiable applications

$$f : \mathcal{M} \rightarrow \mathbb{R}, \qquad (1.80)$$

and with v satisfying the following two conditions:

1. $v(a f + b g) = a v(f) + b v(g)$
2. $v(f \circ g) = v(f)g + f v(g),$ (1.81)

$\forall a, b \in \mathbb{R}, \forall f, g \in \mathcal{F}(\mathcal{M})$. The composed function operator $\circ$ is defined as $(f \circ g)(X) \equiv f(g(X))$. Therefore, given a function $f \in \mathcal{F}(\mathcal{M})$ and a regular point $P \in \mathcal{M}$ we have

$$v_P(f) = v^i(x) \left. \frac{\partial f}{\partial x^i} \right|_P \in \mathbb{R}. \qquad (1.82)$$

1.3.3 Cotangent Space $T_P^*(\mathcal{M})$

The cotangent space is obviously defined as the dual space of the tangent space. Using the same logic as before, we get to the conclusion that a suitable basis for the dual space is formed by the functions dx^i. Therefore, a covector (field) can be written the following way:

$$\boxed{\beta_P = \beta_i(x) \, dx^i \big|_P \in T_P^*(\mathcal{M}); \qquad \beta = \beta_i(x) \, dx^i \in T^*(\mathcal{M})}, \qquad (1.83)$$

where $T^*(\mathcal{M})$ is defined as the union of all $T_P^*(\mathcal{M})$. Using this formalism, given $v \in T_P(\mathcal{M})$ and $\beta \in T_P^*(\mathcal{M})$, $\beta(v)_P \in \mathbb{R}$ is defined as:

$$\boxed{\begin{aligned}
\beta(v)_P &= \beta_j \, dx^j \left(v^i \frac{\partial}{\partial x^i} \right)\bigg|_P \\
&= v^i \, \beta_j \, dx^j \left(\frac{\partial}{\partial x^i} \right)\bigg|_P \\
&= v^i \, \beta_j \, \delta_i^j \big|_P \\
&= v^i(x_P)\beta_i(x_P) \\
&\equiv v(\beta)_P
\end{aligned}} \qquad (1.84)$$

A natural question that might arise is why $dx^j \, (\partial/\partial x^i)|_P = \delta_i^j$? We shall answer this question in a few lines. In differential geometry it is usual to define the following operation. Given $f \in \mathcal{F}(\mathcal{M})$, we define the differential of f evaluated in $P \in \mathcal{M}$ as:

$$\begin{aligned}
df &: T_P(\mathcal{M}) \to \mathbb{R} \\
df &: v_P \mapsto df_P(v) \equiv v_P(f).
\end{aligned} \qquad (1.85)$$

Thus, $df_P(v)$ can be written as

$$df_P(v) = df\left(v^i\frac{\partial}{\partial x^i}\right)\Big|_P = v^i df\left(\frac{\partial}{\partial x^i}\right)\Big|_P \equiv v^i\left(\frac{\partial f}{\partial x^i}\right)\Big|_P. \tag{1.86}$$

Therefore

$$df_P\left(\frac{\partial}{\partial x_j}\right) = df\left(\frac{\partial}{\partial x^j}\right)\Big|_P = \left(\frac{\partial f}{\partial x^j}\right)\Big|_P. \tag{1.87}$$

Taking $df_P = dx^i|_P$, then (1.87) transforms into

$$dx^i\left(\frac{\partial}{\partial x_j}\right)\Big|_P = \left(\frac{\partial x^i}{\partial x^j}\right)\Big|_P = \delta^i{}_j, \tag{1.88}$$

which is exactly the answer we were looking for. As we have seen previously, if v is a vector and β a covector $v(\beta) = \beta(v) = v^i\beta_i$ with $e_j(e^i) \equiv e^i(e_j) = \delta^i{}_j$, therefore it is legitimate to also make the following definition:

$$\boxed{\frac{\partial}{\partial x_j}(dx^i)\Big|_P \equiv dx^i\left(\frac{\partial}{\partial x_j}\right)\Big|_P = \delta^i{}_j.} \tag{1.89}$$

Thus, with this choice of bases we can write any tensor field over $\mathcal{M}$. A few examples are:

$$l = l_{i_1\dots i_m}(x)\, dx^{i_1} \otimes \cdots \otimes dx^{i_m}, \tag{1.90}$$

$$t = t^{i_1\dots i_n}(x)\, \frac{\partial}{\partial x^{i_1}} \otimes \cdots \otimes \frac{\partial}{\partial x^{i_n}}, \tag{1.91}$$

$$q = q_{i_1\dots i_n}{}^{j_1\dots j_m}{}_{k_1\dots k_o}(x) dx^{i_1} \otimes \cdots \otimes dx^{i_n} \otimes \frac{\partial}{\partial x^{j_1}} \otimes \cdots$$
$$\cdots \otimes \frac{\partial}{\partial x^{j_m}} \otimes dx^{k_1} \otimes \cdots \otimes dx^{k_o}, \tag{1.92}$$

etc.

1.3.4 Covariant Derivative

From now on we shall start using the simplified notation: $\partial_i \equiv \dfrac{\partial}{\partial x^i}$ and $\partial'_i \equiv \dfrac{\partial}{\partial x'^i}$. Consider the partial derivative of a scalar field $\partial_j\phi(x)$. Under a change of coordinates $x^i \to x'^i$:

$$\boxed{\partial_i' \phi'(x') = \frac{\partial x^j}{\partial x'^i} \partial_j \phi(x)} . \tag{1.93}$$

We can observe that $\partial_j \phi(x)$ are the components of a rank one covariant tensor field. Let's consider now the derivative of (the components of) a vector field $\partial_i v^j$. Under a change of basis $x^i \to x'^i$, we obtain the following:

$$\partial_i' v'^j = \frac{\partial x^k}{\partial x'^i} \frac{\partial x'^j}{\partial x^l} \partial_k v^l + \frac{\partial x^k}{\partial x'^i} \frac{\partial^2 x'^j}{\partial x^k \partial x^l} v^l . \tag{1.94}$$

Because of the second term on the RHS of (1.94), $\partial_k v^l$ does not behave as a tensor. Can we fix it up? Can we add something else to the *ordinary derivative* ∂_i in order to obtain a tensor quantity? Let's define

$$\boxed{\nabla_i v^j \equiv \partial_i v^j + \Gamma^j_{li} v^l} , \tag{1.95}$$

and try to find the transformation law that the *coefficients* Γ^j_{li} must obey in order to make $\nabla_i v^j$ behave like a $(1, 1)$ tensor:

$$\nabla_i' v'^j = \frac{\partial x^k}{\partial x'^i} \frac{\partial x'^j}{\partial x^l} \nabla_k v^l = \frac{\partial x^k}{\partial x'^i} \frac{\partial x'^j}{\partial x^l} (\partial_k v^l + \Gamma^l_{mk} v^m) = \partial_i' v'^j + \Gamma'^j_{ni} v'^n$$

$$= \frac{\partial x^k}{\partial x'^i} \frac{\partial x'^j}{\partial x^l} \partial_k v^l + \frac{\partial x^k}{\partial x'^i} \frac{\partial^2 x'^j}{\partial x^k \partial x^l} v^l + \frac{\partial x'^n}{\partial x^m} \Gamma'^j_{ni} v^m . \tag{1.96}$$

Therefore, (after changing one mute index) we find

$$\frac{\partial x^k}{\partial x'^i} \frac{\partial x'^j}{\partial x^l} \Gamma^l_{mk} v^m = \frac{\partial x^k}{\partial x'^i} \frac{\partial^2 x'^j}{\partial x^k \partial x^m} v^m + \frac{\partial x'^n}{\partial x^m} \Gamma'^j_{ni} v^m . \tag{1.97}$$

This equality must hold for all v^m, so

$$\frac{\partial x^k}{\partial x'^i} \frac{\partial x'^j}{\partial x^l} \Gamma^l_{mk} = \frac{\partial x^k}{\partial x'^i} \frac{\partial^2 x'^j}{\partial x^k \partial x^m} + \frac{\partial x'^n}{\partial x^m} \Gamma'^j_{ni} . \tag{1.98}$$

Using the fact that:

$$\frac{\partial x^m}{\partial x'^p} \frac{\partial x'^n}{\partial x^m} = \frac{\partial x'^n}{\partial x'^p} = \delta^n_p , \tag{1.99}$$

we find the following transformation rules for Γ'^j_{pi}

$$\boxed{\Gamma'^j_{pi} = \frac{\partial x^m}{\partial x'^p} \frac{\partial x^k}{\partial x'^i} \frac{\partial x'^j}{\partial x^l} \Gamma^l_{mk} - \frac{\partial x^m}{\partial x'^p} \frac{\partial x^k}{\partial x'^i} \frac{\partial^2 x'^j}{\partial x^k \partial x^m}} . \tag{1.100}$$

The symbols Γ^i_{jk} are called the coefficients of the *affine connection* and (1.95) is called the **covariant derivative** of the vector field v. If instead of vector fields we are dealing with covectors, then the covariant derivative takes the following form.

$$\boxed{\nabla_i \beta_j \equiv \partial_i \beta_j - \Gamma^l_{ji} \beta_l}. \tag{1.101}$$

We can extend this to any tensor field. For example, the covariant derivative for a $(2, 2)$ mixed tensor is given by:

$$\boxed{\nabla_i T^{jk}{}_{lm} = \partial_i T^{jk}{}_{lm} + \Gamma^j_{si} T^{sk}{}_{lm} + \Gamma^k_{si} T^{js}{}_{lm} - \Gamma^s_{li} T^{jk}{}_{sm} - \Gamma^s_{mi} T^{jk}{}_{ls}}. \tag{1.102}$$

Taking a quick look at (1.100) we can observe that the quantity $T^i{}_{jk}$ defined as

$$T^i{}_{jk} \equiv \Gamma^i_{jk} - \Gamma^i_{kj}, \tag{1.103}$$

is a $(1, 2)$ anti-symmetric (in the covariant indices) tensor. This tensor is called **torsion tensor**. In General Relativity manifolds are torsion-free.[5] Considering the torsion-free case, the coefficients of the affine connection are symmetric (in the lower indices); they are usually called **Christoffel symbols**. It is easy to demonstrate[6] that they can be expressed in terms of the metric tensor and its first derivatives as:

$$\boxed{\Gamma^i_{jk} = \frac{1}{2} g^{im} \left(\partial_j g_{mk} + \partial_k g_{mj} - \partial_m g_{jk} \right)}. \tag{1.104}$$

The Christoffel symbols are related to the curvature of the manifold. The curvature can be written is terms of the Γ^i_{jk} and its first derivatives. If the manifold is flat, then we can find a global system of coordinates in which all the Christoffel symbols are zero. However, there are system of coordinates for flat manifolds that have non-zero Γ^i_{jk} (i.e., a plane surface expressed in polar coordinates). Curvature of manifolds is a more advanced topic, and its beyond the goal of these notes. The reader is highly encouraged to consult the Further Reading section at the end of the chapter, which are great in treating advanced topics on manifolds, Special and General Relativity.

1.4 Comments on Special Relativity

Before we get to discuss a few Special Relativity topics we need to introduce the Minkowski space. The Minkowski space $\mathcal{M}_4$ is a four dimensional flat manifold.

---

[5]However, there are extensions of the theory which also include torsion.
[6]See Further Reading.

Its tangent space M_4 is a four dimensional vector space together with a metric tensor $g_{\mu\nu} = \text{diag}\{1, -1, -1, -1\}$. Our coordinates in $\mathbb{R}^4$ are the space-time coordinates[7]

$$x^\mu \equiv (x^0, x^1, x^2, x^3) \equiv (t, x^i) \equiv (t, \mathbf{x}). \qquad (1.105)$$

Here we use Greek letters $\mu, \nu = 0, .., 3$ for space-time coordinates and Roman letters $i, j = 1, .., 3$ only for the spatial coordinates. Some authors define x_μ as

$$x_\mu \equiv g_{\mu\nu} x^\nu = (x_0, x_1, x_2, x_3) = (t, x_i) \equiv (t, -x^i) = (t, -\mathbf{x}), \qquad (1.106)$$

but we have to be really careful about that. As we already know x^μ **are coordinates not vector components**, thus x_μ are not covector components either. A well defined vector is dx^μ

$$dx^\mu \equiv (dx^0, dx^1, dx^2, dx^3) \equiv (dt, dx^i) \equiv (dt, d\mathbf{x}). \qquad (1.107)$$

Therefore dx_μ is a properly defined covector

$$dx_\mu \equiv g_{\mu\nu} dx^\nu = (dx^0, -dx^1, -dx^2, -dx^3) \equiv (dt, -dx^i) \equiv (dt, -d\mathbf{x}). \qquad (1.108)$$

After all this being said, the consequences of the postulates of the Special Theory of Relativity can be easily translated mathematically into the following sentence.

The Quantity

$$\boxed{ds^2 = g_{\mu\nu} dx^\mu dx^\nu = dt^2 - d\mathbf{x}^2 = dt^2 - (dx^1)^2 - (dx^2)^2 - (dx^3)^2} \qquad (1.109)$$

must be invariant for any inertial observer. This means

$$ds^2 = g_{\mu\nu} dx^\mu dx^\nu = ds'^2 = g_{\alpha\beta} dx'^\alpha dx'^\beta. \qquad (1.110)$$

Note that on the RHS of the previous equation we haven't written $g'_{\alpha\beta}$, but $g_{\alpha\beta}$ (the metric tensor in Special Relativity does not transform); therefore not all transformations are allowed. The allowed transformations are the ones that maintain invariant

---

[7]The temporal coordinate should really be ct where c is the speed of light in the vacuum, but, as it is usual in Quantum Field Theory, we shall work using natural coordinates $c = 1 = \hbar$.

the metric tensor. These transformations are the ones that belong to the *Poincaré Group*, which can be a *Lorentz Transformation* Λ_{ν}^{μ} plus a *space-time translation* a^{μ} (where a^{μ} are constants)

$$x^{\mu} \rightarrow x'^{\mu} = \Lambda_{\nu}^{\mu} x^{\nu} + a^{\mu}. \tag{1.111}$$

The interesting thing about Lorentz transformation is that they don't depend on x^{μ}:

$$\partial_{\mu} \Lambda_{\beta}^{\alpha} = 0. \tag{1.112}$$

This property allows the partial derivatives to behave as covariant derivatives, therefore we don't have to worry about the Christoffel symbols. However this is only true in **Cartesian coordinates**. If we wanted to work in curvilinear coordinates for example, the property (1.112) wouldn't hold any more. We could generalize everything to general coordinates however, it would be useless in the case of pure Special Relativity. We do this generalization naturally in General Relativity where $g^{\mu\nu}$ is a proper behaved tensor field $g^{\mu\nu} = g^{\mu\nu}(x)$. Returning to our case, taking a quick look at (1.111) and (1.112) we find that the Lorentz transformation is also the contravariant law of transformation for tensors

$$\boxed{\frac{\partial x'^{\mu}}{\partial x^{\nu}} = \Lambda_{\nu}^{\mu}}. \tag{1.113}$$

This allows us to write the equations of motion in a simple, Lorentz-invariant manner (in Cartesian coordinates):

$$\frac{dp^{\mu}}{d\tau} = f^{\mu}, \tag{1.114}$$

where we have defined the momentum four-vector as

$$p^{\mu} = mu^{\mu} = m\frac{dx^{\mu}}{ds} = m\frac{dx^{\mu}}{d\tau} = (m\gamma, m\gamma\mathbf{v}), \tag{1.115}$$

and where $d\tau = ds/c$, the proper time interval (but remember we have set c=1). Any other inertial observer (related to the original one by a Lorentz transformation, or a space-time translation) will describe the equations of motion in the same way

$$\frac{d p'^{\mu}}{d\tau} = f'^{\mu}. \tag{1.116}$$

1.4.1 Lorentz Transformations

Let us quickly review the most important properties of the Lorentz transformations. From (1.110) we obtain

$$g_{\mu\nu}\, dx^\mu\, dx^\nu = g_{\alpha\beta}\, dx'^\alpha\, dx'^\beta$$
$$= g_{\alpha\beta}\, \Lambda^\alpha_\mu\, \Lambda^\beta_\nu\, dx^\mu\, dx^\nu, \qquad (1.117)$$

for arbitrary dx^μ, dx^ν, therefore

$$\boxed{g_{\mu\nu} = g_{\alpha\beta}\, \Lambda^\alpha_\mu\, \Lambda^\beta_\nu}. \qquad (1.118)$$

From the previous equation we straightforwardly obtain

$$(\Lambda^{-1})^\nu_\sigma\, g_{\mu\nu} = g_{\alpha\beta}\, \Lambda^\alpha_\mu\, \Lambda^\beta_\nu\, (\Lambda^{-1})^\nu_\sigma$$
$$= g_{\alpha\beta}\, \Lambda^\alpha_\mu\, \delta^\beta_\sigma$$
$$= g_{\alpha\sigma}\, \Lambda^\alpha_\mu. \qquad (1.119)$$

Contracting with the metric tensor $g^{\mu\rho}$ both sides of the equation, we get to

$$(\Lambda^{-1})^\rho_\sigma = g^{\mu\rho}\, g_{\alpha\sigma}\, \Lambda^\alpha_\mu. \qquad (1.120)$$

This justifies why in QFT literature many authors use the following notation:

$$\boxed{\Lambda^\mu{}_\nu \equiv \Lambda^\mu_\nu, \qquad \Lambda_\nu{}^\mu \equiv (\Lambda^{-1})^\mu_\nu}. \qquad (1.121)$$

With this notation, due to property (1.120) one can relate a Lorentz transformation with its inverse by *effectively* raising and lowering indices **as if** Λ^μ_ν **was** a tensor. We will use this notation from now on.

The relation (1.118) can further give as information on the Lorentz transformations. Using the matrix notation, it reads

$$\Lambda^T g \Lambda = g. \qquad (1.122)$$

Taking the matrix determinant on both sides of the previous equation we obtain

$$[\det(\Lambda)]^2 = 1. \qquad (1.123)$$

Taking into consideration the sign of $\Lambda^0{}_0$ and the value ± 1 of the determinant one can define four sets of Lorentz transformations. A **proper orthochronous** Lorentz transformation (which can be a rotation or a boost)[8] satisfies $\det(\Lambda) = 1$ and $\Lambda^0_{\ 0} \geq 1$. An infinitesimal proper orthochronous Lorentz transformation can be written as a continuous transformation from unity

$$\Lambda^\mu_{\ \nu} = \delta^\mu_\nu + \Delta\omega^\mu_{\ \nu} + \mathcal{O}(\Delta\omega^2), \qquad (1.124)$$

where δ^μ_ν is the Kroneker delta. This expression will turn out to be very useful in the next chapter when we shall introduce the Lagrangian density formalism and Noether's theorem. Looking at (1.118) one can deduce $\Delta\omega^\mu_{\ \nu} = -\Delta\omega^\nu_{\ \mu}$ as follows:

$$g_{\mu\nu} \approx g_{\alpha\beta}\,(\delta^\alpha_\mu + \Delta\omega^\alpha_{\ \mu})\,(\delta^\beta_\nu + \Delta\omega^\beta_{\ \nu})$$
$$\approx g_{\mu\nu} + g_{\alpha\beta}\,\delta^\alpha_\mu\,\Delta\omega^\beta_{\ \nu} + g_{\alpha\beta}\,\delta^\beta_\nu\,\Delta\omega^\alpha_{\ \mu}. \qquad (1.125)$$

Thus $g_{\beta\mu}\,\Delta\omega^\beta_{\ \nu} = -g_{\alpha\nu}\,\Delta\omega^\alpha_{\ \mu}$. Defining

$$\Delta\omega_{\mu\nu} \equiv g_{\mu\alpha}\,\Delta\omega^\alpha_{\ \nu}, \qquad (1.126)$$

we finally obtain $\Delta\omega_{\mu\nu} = -\Delta\omega_{\nu\mu}$, or equivalently $\Delta\omega^\mu_{\ \nu} = -\Delta\omega^\nu_{\ \mu}$, which is the relation we wanted to prove.

Under a proper orthochronous Lorentz transformation the Levi-Civita tensor[9] density (1.72) in four dimensions $\epsilon^{\mu\nu\alpha\beta}$ behaves like a tensor (due to the fact that $\det(\Lambda) = 1$). Other interesting Lorentz transformations (that are not proper orthochronous) can be parity

$$\Lambda^\mu_{\ \nu}(P) = \mathrm{diag}\{1, -1, -1, -1\}, \qquad (1.127)$$

or time reversal

$$\Lambda^\mu_{\ \nu}(T) = \mathrm{diag}\{-1, 1, 1, 1\}. \qquad (1.128)$$

Obviously under these two transformations the Levi-Civita tensor density changes sign.

It is worth mentioning that any arbitrary Lorentz boost (in any arbitrary direction) can be decomposed into rotations and a boost along one axis. Also, any arbitrary Lorentz transformation can be decomposed in terms of boosts, rotations, parity and time reversal.

---

[8]More on boosts and relativistic kinematics will be seen in Chap. 3.
[9]More on the Levi-Civita tensor density in four dimensions will be discussed in Chap. 5.

Further Reading

C.W. Misner, K.S. Thorne y John Archibald Wheeler, *Gravitation*, W. H. Freeman, New York

M.P. Hobson, G.P. Efstathiou, A.N. Lanseby, *General Relativity, An Introduction for Physicists*, Cambridge University Press, Cambridge

J.N. Salas, J.A de Azcarraga, *Class notes*

S. Weinberg, *Gravitation and Cosmology* , Wiley, New York (1972)

A.T. Cantero, M.B. Gambra, *Variedades, tensores y fsica*, http://alqua.tiddlyspace.com/

Y. Choquet-Bruhat, C. De Witt-Morette, M. Dillard-Bleick, *Analysis Manifolds and Physics* (North Holland, 1977)

R. Abraham, J.E. Marsden, T.S. Ratiu, *Manifolds, Tensor Analysis, and Applications*

J.B. Hartle, *Gravity: An Introduction to Einstein's General Relativity*

Further Reading

(faded, illegible text)

Chapter 2
Lagrangians, Hamiltonians and Noether's Theorem

Abstract This chapter is intended to remind the basic notions of the Lagrangian and Hamiltonian formalisms as well as Noether's theorem. We shall first start with a discrete system with N degrees of freedom, state and prove Noether's theorem. Afterwards we shall generalize all the previously introduced notions to continuous systems and prove the generic formulation of Noether's Theorem. Finally we will reproduce a few well known results in Quantum Field Theory.

2.1 Lagragian Formalism

As it is irrelevant for this first part (the discrete case), we shall drop the super-index notation for coordinates or vectors that we have introduced in the previous chapter.

The action associated to a discrete system with N degrees of freedom ($i = 1, \ldots, N$) reads:

$$S(q_i) = \int_{t_1}^{t_2} dt\, L(q_i, \dot{q}_i, t), \qquad (2.1)$$

where $L = L(q_i, \dot{q}_i, t)$ is the Lagrangian of the system and where $\{q_i\}_{i=1}^{N}$ are the generalized coordinates and $\dot{q}_i \equiv dq_i/dt$ the generalized velocities. In order to obtain the Euler-Lagrange equations of motion we consider small variations of the generalized coordinates q_i keeping the extremes fixed:

$$\boxed{q_i' = q_i + \delta q_i, \qquad \delta q_i(t_1) = \delta q_i(t_2) = 0}. \qquad (2.2)$$

The first order Taylor expansion of L then gives

$$L(q_i + \delta q_i, \dot{q}_i + \delta \dot{q}_i, t) = L(q_i, \dot{q}_i, t) + \frac{\partial L}{\partial q_i}\delta q_i + \frac{\partial L}{\partial \dot{q}_i}\delta \dot{q}_i$$

$$\equiv L(q_i, \dot{q}_i, t) + \delta L, \qquad (2.3)$$

© Springer International Publishing Switzerland 2016
V. Ilisie, *Concepts in Quantum Field Theory*,
UNITEXT for Physics, DOI 10.1007/978-3-319-22966-9_2

where summation over repeated indices is also understood. It is straightforward to demonstrate that the *variation* and the *differentiation* operators commute:

$$\delta q_i(t) = q_i'(t) - q_i(t) \;\Rightarrow\; \frac{d}{dt}(\delta q_i) = \dot{q}_i'(t) - \dot{q}_i(t) = \delta \dot{q}_i(t). \qquad (2.4)$$

Thus, we obtain the following expression for δL:

$$\delta L = \frac{\partial L}{\partial q_i}\delta q_i + \frac{\partial L}{\partial \dot{q}_i}\delta \dot{q}_i$$

$$= \frac{\partial L}{\partial q_i}\delta q_i + \frac{\partial L}{\partial \dot{q}_i}\frac{d}{dt}(\delta q_i)$$

$$= \left(\frac{\partial L}{\partial q_i} - \frac{d}{dt}\left(\frac{\partial L}{\partial \dot{q}_i}\right)\right)\delta q_i + \frac{d}{dt}\left(\frac{\partial L}{\partial \dot{q}_i}\delta q_i\right). \qquad (2.5)$$

In order to obtain the equations of motion we apply the **Stationary Action Principle:** *For the physical paths, the action must be a maximum, a minimum or an inflexion point.* This translates mathematically into:

$$\boxed{\delta S = \delta \int_{t_1}^{t_2} dt\, L = \int_{t_1}^{t_2} dt\, \delta L = 0}. \qquad (2.6)$$

Expanding δL we get:

$$\delta S = \int_{t_1}^{t_2} dt \left(\frac{\partial L}{\partial q_i} - \frac{d}{dt}\left(\frac{\partial L}{\partial \dot{q}_i}\right)\right)\delta q_i + \int_{t_1}^{t_2} dt\, \frac{d}{dt}\left(\frac{\partial L}{\partial \dot{q}_i}\delta q_i\right) = 0. \qquad (2.7)$$

Because $\delta q_i(t_1) = \delta q_i(t_2) = 0$ the second integral vanishes:

$$\int_{t_1}^{t_2} dt\, \frac{d}{dt}\left(\frac{\partial L}{\partial \dot{q}_i}\delta q_i\right) = \int_{t_1}^{t_2} d\left(\frac{\partial L}{\partial \dot{q}_i}\delta q_i\right) = \left[\frac{\partial L}{\partial \dot{q}_i}\delta q_i\right]_{t_1}^{t_2} = 0. \qquad (2.8)$$

Therefore, we are left with

$$\delta S = \int_{t_1}^{t_2} dt \left(\frac{\partial L}{\partial q_i} - \frac{d}{dt}\left(\frac{\partial L}{\partial \dot{q}_i}\right)\right)\delta q_i = 0, \qquad (2.9)$$

for arbitrary δq_i. Thus the following equations must hold

$$\boxed{\frac{\partial L}{\partial q_i} - \frac{d}{dt}\left(\frac{\partial L}{\partial \dot{q}_i}\right) = 0}, \qquad (2.10)$$

$\forall q_i$. These equations are called the Euler-Lagrange equations of motion.

From (2.8) we can also deduce an important aspect of Lagrangians, that they are not uniquely defined:

$$\boxed{L(q_i, \dot{q}_i, t) \text{ and } \tilde{L}(q_i, \dot{q}_i, t) = L(q_i, \dot{q}_i, t) + \frac{dF(q_i, t)}{dt}}$$ (2.11)

generate the same equations of motion. We have an alternative way to directly check that adding a function of the form $dF(q_i, t)/dt$ to the Lagrangian, doesn't alter the equations of motion. Applying (2.10) to $dF(q_i, t)/dt$ we obtain:

$$\frac{\partial}{\partial q_i}\left(\frac{dF(q_i, t)}{dt}\right) - \frac{d}{dt}\left(\frac{\partial}{\partial \dot{q}_i}\left(\frac{dF(q_i, t)}{dt}\right)\right) = 0.$$ (2.12)

Next we will present one of the most important theorems of analytical mechanics, a powerful tool that allows us to relate the symmetries of a system with conserved quantities.

2.2 Noether's Theorem

There is a conserved quantity associated with every symmetry of the Lagrangian of a system.

Let's consider a transformation of the type

$$q_i \rightarrow q_i' = q_i + \delta q_i,$$ (2.13)

so that the variation of the Lagrangian can be written as the exact differential of some function F:

$$L(q_i', \dot{q}_i', t) = L(q_i, \dot{q}_i, t) + \frac{dF(q_i, \dot{q}_i, t)}{dt} \implies \delta L = \frac{dF(q_i, \dot{q}_i, t)}{dt}.$$ (2.14)

Note that here we allow F to also depend on $\dot{q}_i$ (that was not the case for (2.11)). On the other hand, we know that we can write δL as:

$$\delta L = \left(\frac{\partial L}{\partial q_i} - \frac{d}{dt}\left(\frac{\partial L}{\partial \dot{q}_i}\right)\right)\delta q_i + \frac{d}{dt}\left(\frac{\partial L}{\partial \dot{q}_i}\delta q_i\right) = \frac{d}{dt}\left(\frac{\partial L}{\partial \dot{q}_i}\delta q_i\right).$$ (2.15)

To get to the last equality we used the equations of motion. Let's now write δq_i as an infinitesimal variation of the form

$$q_i' = q_i + \delta q_i = q_i + \epsilon f_i,$$ (2.16)

with $|\epsilon| \ll 1$ a constant, and f a smooth, well behaved function. Obviously, in the limit $\epsilon \to 0$ we obtain

$$\lim_{\epsilon \to 0} q_i' = q_i \ \Rightarrow \ \lim_{\epsilon \to 0} \delta L = 0. \tag{2.17}$$

Thus, necessarily F must be of the form $F = \epsilon \widetilde{F}$, and

$$\frac{d}{dt}\left(\frac{\partial L}{\partial \dot{q}_i}(\epsilon f_i)\right) = \epsilon \frac{d\widetilde{F}(q_i, \dot{q}_i, t)}{dt}. \tag{2.18}$$

Integrating in t we obtain

$$\frac{\partial L}{\partial \dot{q}_i} f_i = \widetilde{F}(q_i, \dot{q}_i, t) + C, \tag{2.19}$$

with C an integration constant. We therefore conclude, that the conserved quantity associated to our infinitesimal symmetry is:

$$\boxed{C = \frac{\partial L}{\partial \dot{q}_i} f_i - \widetilde{F}(q_i, \dot{q}_i, t)}. \tag{2.20}$$

2.3 Examples

Next, we are going to apply this simple formula to a few interesting cases and reproduce some typical results such as energy and momentum conservation, angular momentum conservation, etc.

2.3.1 Time Translations

Let's consider an infinitesimal time shift: $t \to t + \epsilon$. The first order Taylor expansion of q_i and $\dot{q}_i$ is given by:

$$\delta q_i = q_i(t + \epsilon) - q_i(t) = \epsilon \dot{q}_i(t) + O(\epsilon^2),$$
$$\delta \dot{q}_i = \dot{q}_i(t + \epsilon) - \dot{q}_i(t) = \epsilon \ddot{q}_i(t) + O(\epsilon^2) = \frac{d}{dt}(\delta q_i). \tag{2.21}$$

If the Lagrangian does not exhibit an explicit time dependence ($\partial L/\partial t = 0$) then

$$\delta L = \epsilon \frac{\partial L}{\partial q_i} \dot{q}_i + \epsilon \frac{\partial L}{\partial \dot{q}_i} \ddot{q}_i = \epsilon \frac{dL}{dt} \ \Rightarrow \ \widetilde{F} = L. \tag{2.22}$$

Thus, the conserved quantity is given by the following

$$\boxed{\frac{\partial L}{\partial \dot{q}_i}\dot{q}_i - L = E},$$ (2.23)

where E is the associated energy of the system.

2.3.2 Spatial Translations

Let's consider a Lagrangian of the form $L = T - V$, where T is the kinetic energy of the system and V a central potential. In this case the canonical momentum p_i defined as

$$p_i \equiv \frac{\partial L}{\partial \dot{q}_i},$$ (2.24)

obeys $p_i = \partial T / \partial \dot{q}_i$. Due to the fact that the potential is central and $T \neq T(q_i)$ the Lagrangian obeys

$$L(\mathbf{r}_\alpha + \epsilon \mathbf{n}, \mathbf{v}_\alpha) = L(\mathbf{r}_\alpha, \mathbf{v}_\alpha),$$ (2.25)

with $\mathbf{r}_\alpha$ the coordinates of the particle α and $\mathbf{n}$ an arbitrary spatial direction with $|\mathbf{n}| = 1$. We conclude that $\delta L = 0$. Under this spatial translation the coordinates of the particle α transform the following way:

$$\mathbf{r}_\alpha \to \mathbf{r}'_\alpha = \mathbf{r}_\alpha + \epsilon \mathbf{n},$$ (2.26)

that is

$$r_{\alpha j} \to r'_{\alpha j} = r_{\alpha j} + \epsilon n_j,$$ (2.27)

with $j = 1, 2, 3$. Therefore $f_j = n_j$. The conserved quantity is straightforwardly obtained

$$\boxed{C = \sum_\alpha \frac{\partial L}{\partial \dot{q}_{\alpha j}} n_j = \sum_\alpha p_{\alpha j} n_j = \sum_\alpha \mathbf{p}_\alpha \mathbf{n} = \mathbf{P}\mathbf{n}},$$ (2.28)

for an arbitrary $\mathbf{n}$. Thus, the constant associated to this transformations is the total momentum $\mathbf{P}$ of the system.

2.3.3 Rotations

Again, let's consider a Lagrangian with the same properties as in the previous example. Under an infinitesimal rotation we have

$$\mathbf{r}'_\alpha = \mathbf{r}_\alpha - \epsilon \mathbf{n} \times \mathbf{r}_\alpha, \qquad r'_{\alpha j} = r_{\alpha j} + \epsilon \epsilon_{jkm} n_m r_{\alpha k}, \qquad (2.29)$$

and just as previously $\delta L = 0$. It is straightforward to observe that $f_j = \epsilon_{jkm} n_m r_{\alpha k}$ (where ϵ_{jkm} is the totally antisymmetric three-dimensional Levi-Civita tensor density). The conserved quantity is therefore (remember that summation over all repeated indices is understood):

$$\boxed{C = \frac{\partial L}{\partial \dot{q}_{\alpha j}} \epsilon_{jkm} n_m r_{\alpha k} = p_{\alpha j} \epsilon_{jkm} n_m r_{\alpha k} = (\mathbf{p}_\alpha \times \mathbf{r}_\alpha)\mathbf{n} = -\mathbf{L}\mathbf{n}}. \qquad (2.30)$$

Again, this holds for an arbitrary $\mathbf{n}$, thus, the conserved quantity is the total angular momentum $\mathbf{L}$ of the system.

2.3.4 Galileo Transformations

For this last example we shall consider the same type of Lagrangian as in the previous cases. A Galileo transformation reads

$$\mathbf{r}_\alpha \to \mathbf{r}'_\alpha = \mathbf{r}_\alpha + \mathbf{v}t, \qquad (2.31)$$

with $\mathbf{v}$ a constant velocity vector, therefore:

$$\dot{\mathbf{r}}_\alpha \to \dot{\mathbf{r}}'_\alpha = \dot{\mathbf{r}}_\alpha + \mathbf{v}. \qquad (2.32)$$

Under these transformations $\delta L = \delta T$. Let's calculate T' explicitly:

$$T' = \frac{1}{2} m_\alpha (\dot{\mathbf{r}}_\alpha + \mathbf{v})^2 = T + m_\alpha \dot{\mathbf{r}}_\alpha \mathbf{v} + \sum_\alpha \frac{1}{2} m_\alpha \mathbf{v}^2$$

$$= T + \frac{1}{2} M \mathbf{v}^2 + \frac{d}{dt}(m_\alpha \mathbf{r}_\alpha \mathbf{v}) = T + \frac{1}{2} M \mathbf{v}^2 + \frac{d}{dt}(M \mathbf{R} \mathbf{v}). \qquad (2.33)$$

Considering an infinitesimal transformation $\mathbf{v} = \epsilon \mathbf{n}$ with $|\epsilon| \ll 1$ and ignoring terms of $O(\epsilon^2)$ we have

$$\delta L = \delta T = \epsilon \frac{d}{dt}(M \mathbf{R} \mathbf{n}). \qquad (2.34)$$

The conserved quantity is then given by:

$$C = \sum_{\alpha} p_{\alpha j} n_j t - M\mathbf{R}\mathbf{n} = (\mathbf{P}t - M\mathbf{R})\mathbf{n}, \qquad (2.35)$$

for an arbitrary $\mathbf{n}$. The conserved quantity associated to this transformation is then $\mathbf{P}t - M\mathbf{R}$.

2.4 Hamiltonian Formalism

We define the Hamiltonian functional of a physical system as

$$H(q_i, p_i, t) \equiv p_i \dot{q}_i - L, \qquad (2.36)$$

where p_i is called the canonical conjugated momentum

$$p_i \equiv \frac{\partial L}{\partial \dot{q}_i}, \qquad (2.37)$$

as it was already introduced in (2.24). If the Euler-Lagrange equations (2.10) are satisfied then:

$$\dot{p}_i = \frac{\partial L}{\partial q_i}. \qquad (2.38)$$

The Hamiltonian equations of motion are obtained just as before by applying the principle of the stationary action:

$$\delta S = \int_{t_1}^{t_2} dt \, \delta L$$

$$= \int_{t_1}^{t_2} dt \, \delta(p_i \dot{q}_i - H)$$

$$= \int_{t_1}^{t_2} dt \left(\delta p_i \, \dot{q}_i + p_i \, \delta \dot{q}_i - \frac{\partial H}{\partial q_i} \delta q_i - \frac{\partial H}{\partial p_i} \delta p_i \right)$$

$$= \int_{t_1}^{t_2} dt \left(\delta p_i \, \dot{q}_i + \frac{d}{dt}(p_i \, \delta q_i) - \dot{p}_i \, \delta q_i - \frac{\partial H}{\partial q_i} \delta q_i - \frac{\partial H}{\partial p_i} \delta p_i \right)$$

$$= \int_{t_1}^{t_2} dt \left(\delta p_i \left[\dot{q}_i - \frac{\partial H}{\partial p_i} \right] + \delta q_i \left[-\dot{p}_i - \frac{\partial H}{\partial q_i} \right] \right) + \int_{t_1}^{t_2} d(p_i \, \delta q_i)$$

$$= \int_{t_1}^{t_2} dt \left(\delta p_i \left[\dot{q}_i - \frac{\partial H}{\partial p_i} \right] + \delta q_i \left[-\dot{p}_i - \frac{\partial H}{\partial q_i} \right] \right) = 0. \tag{2.39}$$

This must hold for arbitrary δp_i y δq_i, therefore, the Hamiltonian equations of motion are simply given by:

$$\boxed{\dot{q}_i = \frac{\partial H}{\partial p_i}, \qquad \dot{p}_i = -\frac{\partial H}{\partial q_i}.} \tag{2.40}$$

If the Hamiltonian exhibits an explicit time dependence, it can be easily related to the time dependence of the Lagrangian

$$\frac{dH}{dt} = \frac{d}{dt}(p_i \dot{q}_i - L)$$

$$= \frac{\partial H}{\partial p_i}\dot{p}_i + \frac{\partial H}{\partial q_i}\dot{q}_i + \frac{\partial H}{\partial t}$$

$$= \dot{q}_i \dot{p}_i - \dot{p}_i \dot{q}_i + \frac{\partial H}{\partial t}$$

$$= \dot{p}_i \dot{q}_i + p_i \ddot{q}_i - \frac{\partial L}{\partial q_i}\dot{q}_i - \frac{\partial L}{\partial \dot{q}_i}\ddot{q}_i - \frac{\partial L}{\partial t}. \tag{2.41}$$

Therefore we get to the following simple relation in partial derivatives:

$$\boxed{\frac{\partial H}{\partial t} = -\frac{\partial L}{\partial t}.} \tag{2.42}$$

2.5 Continuous Systems

Until now we have considered discrete systems characterized by N (finite) degrees of freedom. Let's consider now that the system depends on an infinite number of degrees of freedom $N \to \infty$. It no longer makes any sense to talk about discrete coordinates q_i. Instead we have to replace them by a continuous field that is defined for every point in space and that can also vary with time

$$q_i(t) \qquad \to \qquad \phi(\mathbf{x}, t) \equiv \phi(x^\mu) \equiv \phi(x). \tag{2.43}$$

Because now we also have spatial dependence besides time dependence, the following replacement is also justified:

$$\dot{q}_i(t) \quad \rightarrow \quad \left(\partial_t \phi(x), \partial_k \phi(x) \right) \equiv \partial_\mu \phi(x). \tag{2.44}$$

Notice that we have introduced the compact relativistic notation (from Chap. 1) and we have supposed that the partial derivatives of the fields are a Lorentz (or Poincaré) covariant quantity (of the form $\partial_\mu \phi(x)$),[1] with

$$\boxed{\partial_\mu \equiv \frac{\partial}{\partial x^\mu} = (\partial_t, \mathbf{\nabla}) \equiv (\partial_t, \partial_k)}. \tag{2.45}$$

It will also be useful to define the following contravariant quantity

$$\boxed{\partial^\mu \equiv g^{\mu\nu} \partial_\nu = (\partial_t, -\mathbf{\nabla}) \equiv (\partial_t, -\partial_k)}. \tag{2.46}$$

Also, we are only interested in Lagrangians that are invariant under space-time translations besides Lorentz transformations (Poincaré group), therefore they cannot depend explicitly on x^μ. The most generic Lagrangian that exhibits all the properties we have just described can be written as:

$$L = \int_V d^3x \, \mathcal{L}\left(\phi_i(x), \partial_\mu \phi_i(x) \right). \tag{2.47}$$

where $\mathcal{L}$ is called a Lagrangian density (which we will shortly end up calling Lagrangian). Because a system can depend in general on more then one field, we have written our Lagrangian density as a functional of M (with M finite) fields $\{\phi_i\}_{i=1}^M$. Thus, the action can simply be written as an integral of the Lagrangian density

$$S = \int_{t_1}^{t_2} dt \, L$$

$$= \int_{t_1}^{t_2} dt \int_V d^3x \, \mathcal{L}\left(\phi_i(x), \partial_\mu \phi_i(x) \right)$$

$$= \int_{x_1}^{x_2} d^4x \, \mathcal{L}\left(\phi_i(x), \partial_\mu \phi_i(x) \right). \tag{2.48}$$

Just as in the discrete case, in order to obtain the Euler-Lagrange equations of motion we will consider small variations of the fields, keeping the extremes fixed

$$\boxed{\phi'(x) = \phi_i(x) + \delta\phi_i(x); \qquad \delta\phi_i(x_1) = \delta\phi_i(x_2) = 0}. \tag{2.49}$$

---

[1]This is not the most general case, of course, but as we are interested in applying field theory to Special Relativity we shall only restrict our study to this case.

Under these variations, we define

$$\delta\mathcal{L}\Big(\phi_i(x), \partial_\mu\phi_i(x)\Big) \equiv \mathcal{L}\Big(\phi_i(x) + \delta\phi_i(x), \partial_\mu\phi_i(x) + \delta[\partial_\mu\phi_i(x)]\Big)$$
$$- \mathcal{L}\Big(\phi_i(x), \partial_\mu\phi_i(x)\Big), \qquad (2.50)$$

thus, we obtain the following

$$\delta\mathcal{L} = \frac{\partial\mathcal{L}}{\partial\phi_i(x)}\delta\phi_i(x) + \frac{\partial\mathcal{L}}{\partial[\partial_\mu\phi_i(x)]}\delta[\partial_\mu\phi_i(x)]$$

$$= \frac{\partial\mathcal{L}}{\partial\phi_i(x)}\delta\phi_i(x) + \frac{\partial\mathcal{L}}{\partial[\partial_\mu\phi_i(x)]}\partial_\mu[\delta\phi_i(x)]$$

$$= \left(\frac{\partial\mathcal{L}}{\partial\phi_i(x)} - \partial_\mu\frac{\partial\mathcal{L}}{\partial[\partial_\mu\phi_i(x)]}\right)\delta\phi_i(x) + \partial_\mu\left(\frac{\partial\mathcal{L}}{\partial[\partial_\mu\phi_i(x)]}\delta\phi_i(x)\right), \quad (2.51)$$

where summation over all repeated indices is understood. Similar to (2.4), the *variation* and *derivation* operators commute. Applying the principle of the Stationary Action we obtain the Euler-Lagrange equations for continuous systems as follows:

$$\delta S = \int_{x_1}^{x_2}\left(\frac{\partial\mathcal{L}}{\partial\phi_i(x)} - \partial_\mu\frac{\partial\mathcal{L}}{\partial[\partial_\mu\phi_i(x)]}\right)\delta\phi_i(x) + \int_{x_1}^{x_2}\partial_\mu\left(\frac{\partial\mathcal{L}}{\partial[\partial_\mu\phi_i(x)]}\delta\phi_i(x)\right)$$

$$= \int_{x_1}^{x_2}\left(\frac{\partial\mathcal{L}}{\partial\phi_i(x)} - \partial_\mu\frac{\partial\mathcal{L}}{\partial[\partial_\mu\phi_i(x)]}\right)\delta\phi_i(x) = 0, \qquad (2.52)$$

for arbitrary $\delta\phi_i(x)$, therefore, the equations we are looking for take the form

$$\boxed{\frac{\partial\mathcal{L}}{\partial\phi_i(x)} - \partial_\mu\frac{\partial\mathcal{L}}{\partial[\partial_\mu\phi_i(x)]} = 0}, \qquad (2.53)$$

$\forall\,\phi_i$, $i = 1, \ldots, M$. Let's now take another look at (2.52). Because $\delta\phi_i(x_1) = \delta\phi_i(x_2) = 0$, we have found that

$$\int_{x_1}^{x_2}\partial_\mu\left(\frac{\partial\mathcal{L}}{\partial[\partial_\mu\phi_i(x)]}\delta\phi_i(x)\right) = 0. \qquad (2.54)$$

Thus, if we consider an arbitrary functional of the form $b^\mu\Big(\phi_i(x)\Big)$, then

$$\delta\int_{x_1}^{x_2}\partial_\mu b^\mu\Big(\phi_i(x)\Big) = \int_{x_1}^{x_2}\partial_\mu\left(\frac{\partial b^\mu}{\partial\phi_i(x)}\delta[\phi_i(x)]\right) = 0. \qquad (2.55)$$

We conclude that a Lagrangian density is not uniquely defined. Similar to the discrete case, one can always add a functional of the form $\partial_\mu b^\mu \big(\phi_i(x) \big)$ without altering the equations of motion. Therefore

$$
\boxed{\mathcal{L}\big(\phi_i(x),\, \partial_\mu \phi_i(x)\big) \quad \text{and} \quad \mathcal{L}\big(\phi_i(x),\, \partial_\mu \phi_i(x)\big) + \partial_\mu b^\mu \big(\phi_i(x)\big)}, \tag{2.56}
$$

render the same equations of motion.

2.6 Hamiltonian Formalism

We define the Hamiltonian density as

$$
\mathcal{H}\left(\pi_i(x),\, \phi_i(x),\, \nabla \phi_i(x)\right) \equiv \dot{\phi}_i(x)\pi_i(x) - \mathcal{L}, \tag{2.57}
$$

where $\dot{\phi}_i(x) \equiv \partial_t \phi_i(x)$ and $\pi_i(x)$ is the canonical momentum associated to the field $\phi_i(x)$:

$$
\pi_i(x) \equiv \frac{\partial \mathcal{L}}{\partial \dot{\phi}_i(x)}. \tag{2.58}
$$

The action can be written in terms of the Hamiltonian density as

$$
S = \int_{x_1}^{x_2} d^4x\, \mathcal{L} = \int_{x_1}^{x_2} d^4x \left(\dot{\phi}_i(x)\pi_i(x) - \mathcal{H} \right). \tag{2.59}
$$

When applying the principle of the stationary action we obtain

$$
\delta S = \int_{x_1}^{x_2} d^4x \left[\delta\pi_i \left\{ \dot{\phi}_i - \frac{\partial \mathcal{H}}{\partial \pi_i} \right\} - \delta\phi_i \left\{ \dot{\pi}_i + \frac{\partial \mathcal{H}}{\partial \phi_i} - \partial_k \frac{\partial \mathcal{H}}{\partial(\partial_k \phi_i)} \right\} \right] = 0, \tag{2.60}
$$

for arbitrary $\delta\pi_i$ and $\delta\phi_i$. Thus the equations of motion simply read

$$
\boxed{\dot{\phi}(x) = \frac{\partial \mathcal{H}}{\partial \pi(x)}, \qquad \dot{\pi}(x) = -\frac{\partial \mathcal{H}}{\partial \phi(x)} + \partial_k \frac{\partial \mathcal{H}}{\partial(\partial_k \phi(x))}}. \tag{2.61}
$$

where ∂_k are the spatial derivatives ($k = 1, 2, 3$).

2.7 Noether's Theorem (The General Formulation)

Until now we have only introduced a global variation of a field, which is defined as the variation of the shape of the field without changing the space-time coordinates x^μ:

$$\boxed{\delta\phi_i(x) \equiv \phi_i'(x) - \phi_i(x)}. \tag{2.62}$$

Besides this, we can define another type of variation which is closely related, a local variation. It is defined as the difference between the fields evaluated in the same space-time point but in two different coordinates systems:

$$\boxed{\bar{\delta}\phi_i(x) \equiv \phi_i'(x') - \phi_i(x)}. \tag{2.63}$$

Let's now consider a continuous space-time transformation of the type

$$\boxed{x^\mu \to x'^\mu = x^\mu + \Delta x^\mu}, \tag{2.64}$$

which can be a proper orthochronous Lorentz transformation or a space-time translation.[2] At first order in Δx, $\bar{\delta}\phi_i(x)$ reads:

$$\begin{aligned}
\bar{\delta}\phi_i(x) &= \phi_i'(x') - \phi_i(x) \\
&= \phi_i'(x + \Delta x) - \phi_i(x) \\
&\approx \phi_i'(x) + \left(\partial_\mu \phi_i'(x)\right)\Delta x^\mu - \phi_i(x) \\
&\approx \phi_i'(x) + \left(\partial_\mu \phi_i(x)\right)\Delta x^\mu - \phi_i(x) \\
&= \delta\phi_i(x) + \left(\partial_\mu \phi_i(x)\right)\Delta x^\mu.
\end{aligned} \tag{2.65}$$

We therefore, have found the following relation between $\delta\phi(x)$ and $\bar{\delta}\phi(x)$ for an infinitesimal transformation of the type (2.64):

$$\boxed{\bar{\delta}\phi_i(x) = \delta\phi_i(x) + \left(\partial_\mu \phi_i(x)\right)\Delta x^\mu}. \tag{2.66}$$

We can draw the following conclusion. If $\phi_i'(x') = \phi_i(x)$ (which is in general the case for a scalar field; it is also the case for spinor fields under space-time translations) then

$$\delta\phi_i(x) = -\left(\partial_\mu \phi_i(x)\right)\Delta x^\mu. \tag{2.67}$$

---

[2] See Chap. 1 for details.

thus, in this case, an equivalent way of making a transformation of the type (2.64), which acts on the coordinates, is by making an *opposite* transformation on the field:

$$\phi_i(x) \; \rightarrow \; \phi_i'(x) = \phi_i(x - \Delta x). \tag{2.68}$$

Let us now deduce how the Lagrangian transforms under these type of variations. In order to keep the notation short, we shall introduce the following short-hand notations:

$$\mathcal{L}(x) \equiv \mathcal{L}\Big(\phi_i(x), \partial_\mu \phi_i(x)\Big), \qquad \mathcal{L}'(x) \equiv \mathcal{L}\Big(\phi_i'(x), \partial_\mu \phi_i'(x)\Big),$$

$$\mathcal{L}'(x') \equiv \mathcal{L}\Big(\phi_i'(x'), \partial_\mu' \phi_i'(x')\Big), \qquad b^\mu(x) \equiv b^\mu\Big(\phi(x)\Big), \tag{2.69}$$

where $\partial_\mu' \equiv \dfrac{\partial}{\partial x'^\mu}$. Keeping only terms up to $O(\Delta x)$ we can calculate $\bar{\delta}\mathcal{L}(x)$ under (2.64):

$$\bar{\delta}\mathcal{L}(x) = \mathcal{L}'(x') - \mathcal{L}(x)$$

$$= \mathcal{L}\Big(\phi_i(x) + \bar{\delta}\phi_i(x), \partial_\mu \phi_i(x) + \bar{\delta}[\partial_\mu \phi_i(x)]\Big) - \mathcal{L}(x)$$

$$\approx \mathcal{L}(x) + \frac{\partial \mathcal{L}(x)}{\partial \phi_i(x)}\bar{\delta}\phi_i(x) + \frac{\partial \mathcal{L}(x)}{\partial [\partial_\mu \phi_i(x)]}\bar{\delta}[\partial_\mu \phi_i(x)] - \mathcal{L}(x)$$

$$\approx \delta\mathcal{L}(x) + \Big(\partial_\mu \mathcal{L}(x)\Big)\Delta x^\mu, \tag{2.70}$$

where we have introduced the following notation

$$\partial_\mu \mathcal{L}(x) \equiv \frac{\partial \mathcal{L}(x)}{\partial \phi_i(x)}\partial_\mu \phi_i(x) + \frac{\partial \mathcal{L}(x)}{\partial [\partial_\nu \phi_i(x)]}\partial_\mu \partial_\nu \phi_i(x). \tag{2.71}$$

Also we have used the following approximation:

$$\bar{\delta}[\partial_\mu \phi_i(x)] \approx \partial_\mu [\bar{\delta}\phi_i(x)] - \Big(\partial_\nu \phi_i(x)\Big)\frac{\partial \Delta x^\nu}{\partial x^\mu} = \partial_\mu [\bar{\delta}\phi_i(x)]. \tag{2.72}$$

For the last equality we have used that for a Lorentz (Poincaré) transformation $\partial_\mu \Delta x^\nu = 0$ (as it was already mentioned in Chap. 1). Thus, the new *variation* operator $\bar{\delta}$ also commutes with the *derivation* operator when restricting ourselves to Lorentz (Poincaré) continuous transformations. We have therefore obtained an expression similar to (2.66) for $\bar{\delta}\mathcal{L}$:

$$\boxed{\bar{\delta}\mathcal{L}(x) = \delta\mathcal{L}(x) + \Big(\partial_\mu \mathcal{L}(x)\Big)\Delta x^\mu}. \tag{2.73}$$

Let's now consider the transformation of the action. A transformation that leaves the equations of motion invariant is a symmetry of the system. Under such a symmetry the action S will mostly transform as $S \rightarrow S'$ with S' given by

$$
\begin{aligned}
S' &= \int_{\Omega'} d^4x' \mathcal{L}'(x') \\
&= \int_{\Omega} d^4x \mathcal{L}(x) + \int_{\Omega} d^4x \, \partial_\mu b^\mu(x) \\
&= S + \int_{\Omega} d^4x \, \partial_\mu b^\mu(x),
\end{aligned}
\tag{2.74}
$$

so that $\delta S' = \delta S$ (thus generating the same equations of motion). Introducing the Jacobian matrix we have

$$
\int_{\Omega} \left| \frac{\partial x'}{\partial x} \right| d^4x \mathcal{L}'(x') = \int_{\Omega} d^4x \mathcal{L}(x) + \int_{\Omega} d^4x \, \partial_\mu b^\mu(x).
\tag{2.75}
$$

This must hold for all space-time volumes Ω, therefore:

$$
\left| \frac{\partial x'}{\partial x} \right| \mathcal{L}'(x') = \mathcal{L}(x) + \partial_\mu b^\mu(x).
\tag{2.76}
$$

The determinant of the Jacobian matrix is equal to 1 for a proper orthocronous Lorentz transformation or a space-time translation, thus

$$
\boxed{ \bar{\delta} \mathcal{L}(x) - \partial_\mu b^\mu(x) = 0 }.
\tag{2.77}
$$

Introducing (2.73) in (2.77) we obtain:

$$
\boxed{ \delta \mathcal{L}(x) + \partial_\mu \left[\mathcal{L}(x) \Delta x^\mu - b^\mu(x) \right] = 0 }.
\tag{2.78}
$$

Inserting the explicit form of $\delta \mathcal{L}$ from (2.51) in the last expression, we obtain

$$
\left(\frac{\partial \mathcal{L}(x)}{\partial \phi_i(x)} - \partial_\mu \frac{\partial \mathcal{L}(x)}{\partial [\partial_\mu \phi_i(x)]} \right) \delta \phi_i(x) +
$$
$$
+ \partial_\mu \left(\frac{\partial \mathcal{L}(x)}{\partial [\partial_\mu \phi_i(x)]} \delta \phi_i(x) \right) + \partial_\mu \left[\mathcal{L}(x) \Delta x^\mu - \partial_\mu b^\mu(x) \right] = 0.
\tag{2.79}
$$

Using the Euler-Lagrange equations of motion we finally get to the conservation law we were looking for

$$
\boxed{ \partial_\mu j^\mu(x) = 0, \qquad j^\mu(x) = \frac{\partial \mathcal{L}(x)}{\partial [\partial_\mu \phi_i(x)]} \delta \phi_i(x) + \mathcal{L}(x) \Delta x^\mu - b^\mu(x) },
\tag{2.80}
$$

with $j^\mu(x)$ the conserved Noether current. Note that our result is completely general, in the sense that it holds for continuous space-time transformations of the type (2.64) and also for transformations that only imply a field variation without modifying the space-time configuration. In this last case we would simply set $\Delta x^\mu = 0$ in (2.80). As it is usual, we can also define a **conserved charge** $\mathcal{Q}$ associated to the conserved current j^μ as:

$$\boxed{\mathcal{Q} = \int d^3x\, j^0, \qquad \frac{d\mathcal{Q}}{dt} = \int d^3x\, \partial_0 j^0 = -\int d^3x\, \nabla \mathbf{j} = 0}. \qquad (2.81)$$

Next, we shall take a few illustrative examples.

2.8 Examples

2.8.1 Space-time Translations

Consider the following infinitesimal space-time translation:

$$x^\mu \rightarrow x'^\mu = x^\mu - \epsilon^\mu, \qquad (2.82)$$

with ϵ^μ real constants. For scalar or spinor fields we have $\phi_i(x) = \phi_i'(x')$ thus $\bar{\delta}\phi_i(x) = 0$. Under this type of transformation our Lagrangians remain unchanged so $\mathcal{L}'(x') = \mathcal{L}(x)$, therefore, by taking a look at (2.77) we conclude that $\partial_\mu b^\mu = 0$. We can thus, eliminate the b^μ term from (2.80) and the conserved current is simply given by:

$$j^\mu = \frac{\partial \mathcal{L}}{\partial(\partial_\mu \phi_i)} \partial_\nu \phi_i\, \epsilon^\nu - \mathcal{L}\epsilon^\mu = \left(\frac{\partial \mathcal{L}}{\partial(\partial_\mu \phi_i)} \partial^\nu \phi_i - \mathcal{L} g^{\mu\nu} \right) \epsilon_\nu. \qquad (2.83)$$

The conservation law $\partial_\mu j^\mu = 0$ holds for any arbitrary constants ϵ_ν, therefore we actually have four conserved currents:

$$\partial_\mu T^{\mu\nu} = 0, \qquad T^{\mu\nu} = \frac{\partial \mathcal{L}}{\partial(\partial_\mu \phi_i)} \partial^\nu \phi_i - \mathcal{L} g^{\mu\nu}, \qquad (2.84)$$

with $T^{\mu\nu}$ the four-momentum tensor. The conserved Noether charges are then given by

$$\mathcal{P}^\nu \equiv \int d^3x\, T^{0\nu}$$

$$= \int d^3x \left(\frac{\partial \mathcal{L}}{\partial \dot{\phi}_i} \partial^\nu \phi_i - g^{0\nu} \mathcal{L} \right)$$

$$= \int d^3x \, (\pi_i \, \partial^\nu \phi_i - g^{0\nu} \mathcal{L})$$

$$= \int d^3x \, (\pi_i \, \dot{\phi}_i - g^{00} \mathcal{L}, \, -\pi_i \, \nabla \phi_i)$$

$$= \int d^3x \, (\mathcal{H}, \mathcal{P})$$

$$= (H, \mathbf{P}), \tag{2.85}$$

where we have used (2.46). As we can see, the conserved charges are the Hamiltonian and three-momentum operators.

2.8.2 Phase Redefinition

Consider a Lagrangian that depends on the fields ϕ_1 and ϕ_2 with $\phi_1 = \phi$ and $\phi_2 = \phi^\dagger$. If we perform an infinitesimal global phase redefinition of the field

$$\phi(x) \rightarrow \phi'(x) = e^{-i\theta} \phi(x), \tag{2.86}$$

with $\theta \ll 1$ (and where global means that the phase does not depend on the space-time coordinates $\theta \neq \theta(x)$), then we find:

$$\delta\phi(x) = -i\theta\phi(x), \qquad \delta\phi^\dagger(x) = i\theta\phi^\dagger(x). \tag{2.87}$$

As this transformation doesn't involve the space-time coordinates we can already set $\Delta x^\mu = 0$ in (2.80). Therefore $\delta\mathcal{L} = \bar{\delta}\mathcal{L} = \partial_\mu b^\mu$. Again, if we only consider the free Dirac or Klein-Gordon Lagrangians then $\delta\mathcal{L} = 0 = \partial_\mu b^\mu$, so we can also eliminate b^μ from (2.80). The conserved current is then given by

$$j^\mu = \frac{\partial \mathcal{L}}{\partial(\partial_\mu \phi)} \delta\phi + \frac{\partial \mathcal{L}}{\partial(\partial_\mu \phi^\dagger)} \delta\phi^\dagger = \frac{\partial \mathcal{L}}{\partial(\partial_\mu \phi)} (-i\phi)\theta + \frac{\partial \mathcal{L}}{\partial(\partial_\mu \phi^\dagger)} i\phi^\dagger \theta, \tag{2.88}$$

for an arbitrary θ. Thus, redefining the current without the θ multiplying term we find

$$\partial_\mu j^\mu = 0, \qquad j^\mu(x) = -i\frac{\partial \mathcal{L}}{\partial(\partial_\mu \phi)} \phi + i\frac{\partial \mathcal{L}}{\partial(\partial_\mu \phi^\dagger)} \phi^\dagger. \tag{2.89}$$

In particular, for the free Dirac Lagrangian $\mathcal{L}_D = \bar{\psi}(x)(i\gamma_\mu \partial^\mu - m)\psi(x)$ we obtain the well known result:

$$\partial_\mu \left(\bar{\psi}\gamma^\mu\psi \right) = 0. \tag{2.90}$$

2.8.3 Lorentz Transformations

Consider the following infinitesimal proper orthochronous[3] Lorentz transformation:

$$x^\mu \rightarrow x'^\mu = x^\mu + \Delta\omega^\mu_{\ \nu}\, x^\nu, \tag{2.91}$$

with $\Delta\omega^\mu_{\ \nu} = -\Delta\omega^\nu_{\ \mu}$ real constants. Defining $\Delta\omega_{\mu\nu} \equiv g_{\mu\alpha}\,\Delta\omega^\alpha_{\ \nu}$, it is easy to show that the field transformation reads

$$\phi'_i(x') = \phi_i(x) + \frac{1}{2}\Sigma^{\mu\nu}_{(i)}\,\Delta\omega_{\mu\nu}\,\phi_i(x), \tag{2.92}$$

with $\Sigma^{\mu\nu}_{(i)} = -\frac{i}{2}\sigma^{\mu\nu} = \frac{1}{4}[\gamma^\mu, \gamma^\nu]$ for spinorial[4] fields and, $\Sigma^{\mu\nu}_{(i)} = 0$ for scalar fields (no summation over the "i" index must be understood in (2.92) nor in the following expression). Using (2.66) we easily find:

$$\delta\phi_i(x) = \frac{1}{2}\Sigma^{\alpha\beta}_{(i)}\,\Delta\omega_{\alpha\beta}\,\phi_i(x) - \partial^\alpha\phi_i(x)\,\Delta\omega_{\alpha\beta}x^\beta$$

$$= \frac{1}{2}\Sigma^{\alpha\beta}_{(i)}\,\Delta\omega_{\alpha\beta}\,\phi_i(x) - \frac{1}{2}\left(\partial^\alpha\phi_i(x)\,x^\beta - \partial^\beta\phi_i(x)\,x^\alpha\right)\Delta\omega_{\alpha\beta}$$

$$= \frac{1}{2}\left[\Sigma^{\alpha\beta}_{(i)} + \left(x^\alpha\partial^\beta - x^\beta\partial^\alpha\right)\right]\phi_i(x)\,\Delta\omega_{\alpha\beta}. \tag{2.93}$$

On the other hand, our Lagrangians are all Lorentz invariant, thus $\bar\delta\mathcal{L} = 0$ and so again, we can eliminate b^μ in (2.80) just as in the previous examples. We obtain that the expression for our conserved current reads:

$$j^\mu = \frac{\partial\mathcal{L}}{\partial[\partial_\mu\phi_i]}\left(\Sigma^{\alpha\beta}_{(i)} + x^\alpha\partial^\beta - x^\beta\partial^\alpha\right)\phi_i\,\frac{1}{2}\Delta\omega_{\alpha\beta} + \mathcal{L}\,\Delta\omega^\mu_{\ \beta}x^\beta$$

$$= \frac{\partial\mathcal{L}}{\partial[\partial_\mu\phi_i]}\left(\Sigma^{\alpha\beta}_{(i)} + x^\alpha\partial^\beta - x^\beta\partial^\alpha\right)\phi_i\,\frac{1}{2}\Delta\omega_{\alpha\beta} +$$

$$+ \frac{1}{2}\mathcal{L}\,(g^{\mu\alpha}x^\beta - g^{\mu\beta}x^\alpha)\,\Delta\omega_{\alpha\beta}, \tag{2.94}$$

for arbitrary $\Delta\omega_{\alpha\beta}$. Thus, we obtain

$$\partial_\mu \mathcal{J}^{\mu,\alpha\beta} = 0, \qquad \mathcal{J}^{\mu,\alpha\beta} = x^\alpha T^{\mu\beta} - x^\beta T^{\mu\alpha} + \frac{\partial\mathcal{L}}{\partial[\partial_\mu\phi_i]}\Sigma^{\alpha\beta}_{(i)}\phi_i, \tag{2.95}$$

---

[3] See Chap. 1 for more details.
[4] See Chap. 5 for details on spinor algebra and for the proof of this statement.

which is the conservation law of the angular momentum *pseudo*[5] tensor $\mathcal{J}^{\mu,\alpha\beta}$ (obviously, for the previous expression, summation over all repeated indices must be understood).

Further Reading

A. Pich, *Class Notes on Quantum Field Theory*. http://eeemaster.uv.es/course/view.php?id=6

W. Greiner, J. Reinhardt, D.A. Bromley (Foreword), *Field Quantization*

E.L. Hill, Hamilton's principle and the conservation theorems of mathematical physics. Rev. Mod. Phys. **23**, 253

J.A. de Azcárraga, J.M. Izquierdo, *Lie Groups, Lie Algebras, Cohomology and Some Applications in Physics*. Cambridge Monographs in Mathematical Physics

J.A. Oller, Mecnica Terica, http://www.um.es/oller/docencia/versionmteor.pdf

M. Kaksu, *Quantum Field Theory: A Modern Introduction*

M. Srednicki, *Quantum Field Theory*

D.E. Soper, *Classical Field Theory*

D.V. Galtsov, Iu.V. Grats, Ch. Zhukovski, *Campos Clásicos*

S. Noguera, *Class Notes*

---

[5]I am calling it pseudo tensor because it is obviously not invariant under translations!.

Chapter 3
Relativistic Kinematics and Phase Space

Abstract Here we present a list of the most important formulae needed for calculating relativistic collisions and decays. It includes one-to-two and one-to-three body decays, and the two-to-two scattering process both in the center of mass and laboratory frames. It also includes simplified general formulae of one, two and three-body Lorentz invariant phase space. No explicit calculation is performed, however the reader is highly encouraged to reproduce the results presented here.

3.1 Conventions and Notations

For all the calculations in this book we will adopt the *mostly minus* Minkowski metric $g = \text{diag}\{1, -1, -1, -1\}$. If a particle has a relativistic three-momentum $\mathbf{p} = \gamma m \mathbf{v}$, then we define the contravariant four-momentum vector as:

$$p^\mu = (E, \mathbf{p}) = (\gamma m, \gamma m \mathbf{v}). \tag{3.1}$$

Thus $p^2 \equiv p_\mu p^\mu = m^2$; here we have taken $c = 1$ as usual. We will also be needing the Kallen lambda function defined as

$$\lambda(x, y, z) \equiv x^2 + y^2 + z^2 - 2xy - 2yz - 2xz. \tag{3.2}$$

Another important issue is to fix the signs for the Lorentz boosts. Here we will use the passive transformation approach described as follows. Consider two inertial reference frames $\mathcal{O}$ and $\mathcal{O}'$ with all axes parallel and with $\mathcal{O}'$ moving in the positive direction of the $\hat{z}$ axis with constant velocity $\mathbf{v} = v\hat{z}$ ($v > 0$) relative to $\mathcal{O}$ (as shown in Fig. 3.1). Thus, an observer from the reference frame $\mathcal{O}'$ *sees* that $\mathcal{O}$ moves with velocity $-\mathbf{v}$ relative to his reference frame (do not mistake $\mathbf{v}$ of the reference frame with the velocity of the particle from (3.1)). Mathematically this translates into:

$$x'^\mu = \Lambda^\mu_{\ \nu} x^\nu \tag{3.3}$$

© Springer International Publishing Switzerland 2016
V. Ilisie, *Concepts in Quantum Field Theory*,
UNITEXT for Physics, DOI 10.1007/978-3-319-22966-9_3

Fig. 3.1 Two inertial
reference frames $\mathcal{O}$ and $\mathcal{O}'$
with parallel axes and
relative constant velocity

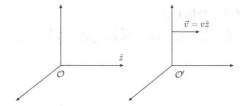

with $\Lambda^{\mu}{}_{\nu}$ given by

$$\Lambda^{\mu}{}_{\nu} = \begin{pmatrix} \gamma & 0 & 0 & -\gamma v \\ 0 & 1 & 0 & 0 \\ 0 & 0 & 1 & 0 \\ -\gamma v & 0 & 0 & \gamma \end{pmatrix}. \tag{3.4}$$

Therefore, a moving object that has four momentum p^{μ} relative to the reference
frame $\mathcal{O}$ will be described as having four momentum $p'^{\mu} = \Lambda^{\mu}{}_{\nu} p^{\nu}$ by an observer
in $\mathcal{O}'$. This will turn out to be useful, for example, when calculating the relation
between the center of mass collision angle and the one in the laboratory reference
frame for the process $a + b \rightarrow 1 + 2$. It will also be needed in Chap. 4, where
we will describe the three and four-body kinematics and phase-space in terms of
angular observables. The inverse transformation is obtained simply by making the
substitution $-\gamma v \rightarrow +\gamma v$ in (3.4).

3.2 Process: $a \rightarrow 1 + 2$

In the center of mass (CM) reference frame we have the following configuration:

$$p_a^{\mu} = (m_a, \mathbf{0}), \qquad p_1^{\mu} = (E_1, -\mathbf{p}), \qquad p_2^{\mu} = (E_2, \mathbf{p}). \tag{3.5}$$

where E_1, E_2 and $|\mathbf{p}|$ are given by

$$E_1 = \frac{1}{2m_a}(m_a^2 + m_1^2 - m_2^2),$$

$$E_2 = \frac{1}{2m_a}(m_a^2 + m_2^2 - m_1^2),$$

$$|\mathbf{p}| = \frac{1}{2m_a}\lambda^{1/2}(m_a^2, m_1^2, m_2^2). \tag{3.6}$$

The threshold value of s (minimum value of s for the on-shell production of particles
1 and 2) is: $s_{th} = (m_1 + m_2)^2$, thus $s \geqslant s_{th}$.

3.3 Process: $a \rightarrow 1 + 2 + 3$

In the CM frame we have the following configuration:

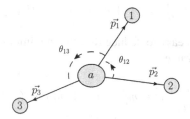

$$p_a^\mu = (m_a, \mathbf{0}), \qquad\qquad p_i^\mu = (E_i, \mathbf{p_i}),$$
$$\sum_i \mathbf{p_i} = \mathbf{0}, \qquad\qquad m_a = \sum_i E_i. \qquad (3.7)$$

Because $\mathbf{p_1} + \mathbf{p_2} + \mathbf{p_3} = \mathbf{0}$ the process takes place in the same plane. Using the angles shown in the previous figure:

$$|\mathbf{p_1}| + |\mathbf{p_2}| \cos\theta_{12} + |\mathbf{p_3}| \cos\theta_{13} = 0,$$
$$|\mathbf{p_2}| \sin\theta_{12} + |\mathbf{p_3}| \sin\theta_{13} = 0. \qquad (3.8)$$

The standard approach for the definition of the Lorentz invariant kinematical variables is given by:

$$t_1 \equiv s_{23} \equiv (p_a - p_1)^2 = (p_2 + p_3)^2,$$
$$t_2 \equiv s_{13} \equiv (p_a - p_2)^2 = (p_1 + p_3)^2,$$
$$t_3 \equiv s_{12} \equiv (p_a - p_3)^2 = (p_1 + p_2)^2. \qquad (3.9)$$

The t_i invariants satisfy the following relation:

$$\sum_i t_i = m_a^2 + \sum_i m_i^2. \qquad (3.10)$$

It is easy to show that they also satisfy:

$$t_i = (p_a - p_i)^2 = p_a^2 + p_i^2 - 2p_a \cdot p_i = m_a^2 + m_i^2 - 2m_a \, E_i. \qquad (3.11)$$

Therefore we can express E_i and $|\mathbf{p_i}|$ in terms of Lorentz invariant quantities as

$$E_i = \frac{1}{2m_a}(m_a^2 + m_i^2 - t_i).$$
$$|\mathbf{p_i}| = \frac{1}{2m_a}\lambda^{1/2}(m_a^2, m_i^2, t_i). \qquad (3.12)$$

The threshold values for s and s_{ij} are denoted as s_{th} and s_{ij}^{th} and:

$$s \equiv p_a^2 = m_a^2 \geqslant s_{th} = (m_1 + m_2 + m_3)^2,$$
$$s_{ij} = t_k \geqslant s_{ij}^{th} = (m_i + m_j)^2 = t_k^{th}, \quad (i \neq j \neq k). \tag{3.13}$$

From (3.12) and (3.13) is easy to deduce that the maximum value of the energy of the particle i in the CM frame is

$$E_i^{\max} = \frac{1}{2m_a}(m_a^2 + m_i^2 - (m_j + m_k)^2), \quad (i \neq j \neq k). \tag{3.14}$$

3.4 Process: $1 + 2 \rightarrow 3 + 4$

In the CM reference frame we have the following configuration:

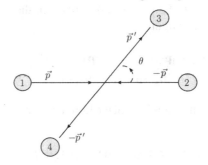

$$\begin{array}{ll} p_1^\mu = (E_1, \mathbf{p}), & p_2^\mu = (E_2, -\mathbf{p}), \qquad (3.15) \\ p_3^\mu = (E_3, \mathbf{p}'), & p_4^\mu = (E_4, -\mathbf{p}'). \qquad (3.16) \end{array}$$

We can define the following Lorentz invariant quantities:

$$s \equiv (p_1 + p_2)^2 = (p_3 + p_4)^2,$$
$$t \equiv (p_1 - p_3)^2 = (p_2 - p_4)^2,$$
$$u \equiv (p_1 - p_4)^2 = (p_2 - p_3)^2. \tag{3.17}$$

One can easily check that:

$$s + t + u = \sum_i m_i^2. \tag{3.18}$$

The CM energies and momenta in terms of s and m_i are given by:

$$E_1 = \frac{1}{2\sqrt{s}}(s + m_1^2 - m_2^2), \qquad E_3 = \frac{1}{2\sqrt{s}}(s + m_3^2 - m_4^2),$$

$$E_2 = \frac{1}{2\sqrt{s}}(s + m_2^2 - m_1^2), \qquad E_4 = \frac{1}{2\sqrt{s}}(s + m_4^2 - m_3^2),$$

$$|\mathbf{p}| = \frac{1}{2\sqrt{s}}\lambda^{1/2}(s, m_1^2, m_2^2), \qquad |\mathbf{p}'| = \frac{1}{2\sqrt{s}}\lambda^{1/2}(s, m_3^2, m_4^2). \tag{3.19}$$

The CM collision angle, shown in the previous figure, in terms of Lorentz invariant quantities is given by

$$\cos\theta = \frac{s(t - u) + (m_1^2 - m_2^2)(m_3^2 - m_4^2)}{\lambda^{1/2}(s, m_1^2, m_2^2)\lambda^{1/2}(s, m_3^2, m_4^2)}. \tag{3.20}$$

In the laboratory (L) reference frame we consider that the particle 2 is at rest, thus we have the following configuration:

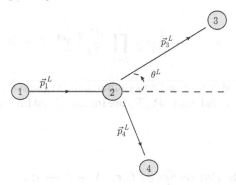

$$p_{1,L}^\mu = (E_1^L, \, \mathbf{p}_1^L), \qquad p_{2,L}^\mu = (m_2, \, \mathbf{0}),$$

$$p_{3,L}^\mu = (E_3^L, \, \mathbf{p}_3^L), \qquad p_{4,L}^\mu = (E_4^L, \, \mathbf{p}_4^L). \tag{3.21}$$

The energies and momenta in the L reference frame are given by:

$$E_1^L = \frac{1}{2m_2}(s - m_1^2 - m_2^2), \qquad |\mathbf{p}_1^L| = \frac{1}{2m_2}\lambda^{1/2}(s, m_2^2, m_1^2),$$

$$E_4^L = \frac{1}{2m_2}(m_2^2 + m_4^2 - t), \qquad |\mathbf{p}_4^L| = \frac{1}{2m_2}\lambda^{1/2}(t, m_2^2, m_4^2),$$

$$E_3^L = \frac{1}{2m_2}(m_2^2 + m_3^2 - u), \qquad |\mathbf{p}_3^L| = \frac{1}{2m_2}\lambda^{1/2}(u, m_2^2, m_3^2). \tag{3.22}$$

The expression for θ^L scattering angle can be easily related to the CM one with the following expression

$$\tan \theta^L = \frac{m_2 |\mathbf{p}'| \sin \theta}{|\mathbf{p}| E_3 + |\mathbf{p}'| E_2 \cos \theta}, \tag{3.23}$$

where all the quantities from the RHS of the equation are expressed in the CM reference frame.

Next we present simple expressions for the Lorentz invariant phase space for one, two and three final state particles in terms of the previously discussed kinematical variables.

3.5 Lorentz Invariant Phase Space

The standard definition for the Lorentz invariant phase space for N final state particles is given by

$$dQ_N \equiv S_n \frac{1}{(2\pi)^{3N-4}} \prod_{l=1}^{N} \frac{d^3 p_l}{2E_l} \delta^{(4)}(\mathcal{P}_i - \mathcal{P}_f), \tag{3.24}$$

where $S_n = 1/n!$ is the corresponding symmetry factor (with n the number of final state identical particles) and where $\mathcal{P}_i$, $\mathcal{P}_f$ are the initial and final state total momenta of the system.

3.5.1 One Particle Phase Space (i.e., $1 + 2 \rightarrow a$)

We have the following simple expression:

$$dQ_1 = 2\pi \frac{d^3 p_a}{2E_a} \delta^{(4)}(\mathcal{P}_i - \mathcal{P}_f). \tag{3.25}$$

We can easily integrate the previous expression and obtain

$$\int dQ_1 = 2\pi \delta(s - m_a). \tag{3.26}$$

3.5.2 Two Particle Phase Space (i.e., $1(+2) \rightarrow 3 + 4$)

We have the following expression:

$$dQ_2 = S_n \frac{1}{(2\pi)^2} \frac{d^3 p_3}{2E_3} \frac{d^3 p_4}{2E_4} \delta^{(4)}(\mathcal{P}_i - \mathcal{P}_f). \tag{3.27}$$

If we choose to work in the CM reference frame we simply obtain:

$$dQ_2 = S_n \frac{1}{(2\pi)^2} \frac{|\mathbf{p}'|}{4\sqrt{s}} d\Omega, \tag{3.28}$$

where $|\mathbf{p}'|$ is the final state CM momentum as in (3.19). The differential solid angle is given by $d\Omega = d\cos\theta\, d\phi$. If the scattering matrix does not depend on the phase space then we can perform the angular integration:

$$\int d\Omega = \int_{-1}^{1} d\cos\theta \int_{0}^{2\pi} d\phi = 4\pi. \tag{3.29}$$

Sometimes is much simpler to express the scattering matrix as a function of s, t and the corresponding masses, and $d\cos\theta$ in terms those same variables. Taking a quick look at (3.18), we find that for a collision at fixed s (which is the most typical case for colliders, thus $ds = 0$) we have

$$dt = -du. \tag{3.30}$$

Therefore, using (3.20) we can express $d\cos\theta$ as:

$$d\cos\theta = \frac{2s\, dt}{\lambda^{1/2}(s, m_1^2, m_2^2)\, \lambda^{1/2}(s, m_3^2, m_4^2)}. \tag{3.31}$$

The integration limits are easy to obtain using the same expression (3.20):

$$(\cos\theta)^{\max} = \frac{s\,(2\,t^{\max} + s - \sum m_i^2) + (m_1^2 - m_2^2)(m_3^2 - m_4^2)}{\lambda^{1/2}(s, m_1^2, m_2^2)\, \lambda^{1/2}(s, m_3^2, m_4^2)} = 1, \tag{3.32}$$

$$(\cos\theta)^{\min} = \frac{s\,(2\,t^{\min} + s - \sum m_i^2) + (m_1^2 - m_2^2)(m_3^2 - m_4^2)}{\lambda^{1/2}(s, m_1^2, m_2^2)\, \lambda^{1/2}(s, m_3^2, m_4^2)} = -1. \tag{3.33}$$

Thus, we simply get:

$$t^{\max} = \frac{1}{2}\sum m_i^2 - \frac{s}{2} - \frac{1}{2s}(m_1^2 - m_2^2)(m_3^2 - m_4^2)$$
$$+ \frac{1}{2s}\lambda^{1/2}(s, m_1^2, m_2^2)\,\lambda^{1/2}(s, m_3^2, m_4^2), \tag{3.34}$$

$$t^{min} = \frac{1}{2} \sum m_i^2 - \frac{s}{2} - \frac{1}{2s}(m_1^2 - m_2^2)(m_3^2 - m_4^2)$$

$$- \frac{1}{2s} \lambda^{1/2}(s, m_1^2, m_2^2) \lambda^{1/2}(s, m_3^2, m_4^2). \tag{3.35}$$

If $m_1 = m_2$ or $m_3 = m_4$ these expressions get a lot simpler. If we are dealing with a decay $1 \to 3 + 4$, then we must set $m_2 = 0$ and $s = m_1^2$.

3.5.3 Three Particle Phase Space (i.e., $a\,(+b) \to 1 + 2 + 3$)

The Lorentz invariant phase space corresponding to three final state particles is given by:

$$dQ_3 = S_n \frac{1}{(2\pi)^5} \frac{d^3 p_1}{2E_1} \frac{d^3 p_2}{2E_2} \frac{d^3 p_3}{2E_3} \delta^{(4)}(\mathcal{P}_i - \mathcal{P}_f). \tag{3.36}$$

In the CM reference frame we obtain the simple and compact expression

$$dQ_3 = S_n \frac{1}{128\,\pi^3\,s} \, ds_{23}\, ds_{13}. \tag{3.37}$$

The integration limits are given by:

$$s_{13}^{min} = (m_1 + m_3)^2, \qquad\qquad s_{13}^{max} = (\sqrt{s} - m_2)^2, \tag{3.38}$$

and

$$s_{23}^{min} = \frac{1}{4s_{13}} \Big\{ (s - m_1^2 - m_2^2 + m_3^2)^2$$

$$- \big[\lambda^{1/2}(s, s_{13}, m_2^2) + \lambda^{1/2}(s_{13}, m_1^2, m_3^2) \big]^2 \Big\}, \tag{3.39}$$

$$s_{23}^{max} = \frac{1}{4s_{13}} \Big\{ (s - m_1^2 - m_2^2 + m_3^2)^2$$

$$- \big[\lambda^{1/2}(s, s_{13}, m_2^2) - \lambda^{1/2}(s_{13}, m_1^2, m_3^2) \big]^2 \Big\}, \tag{3.40}$$

and where we have defined $s \equiv (p_1 + p_2 + p_3)^2 = (p_a + p_b)^2$ as the CM invariant energy (fixed for colliders). If it's a decay i.e., $a \to 1 + 2 + 3$ then we must simply set $s = m_a^2$.

Finally, we will we will provide the decay rate and cross section formulae corresponding to our conventions. If $\mathcal{M}$ is the transition matrix for a given process, the decay rate is given by

$$
\boxed{
\begin{aligned}
\Gamma(a \to 1 + 2 + \cdots + N) &= \frac{1}{2\,m_a} \frac{1}{(2j_a + 1)} \sum_{\lambda_a, \lambda_1, \ldots, \lambda_N} \\
&\times \int dQ_N \left| \mathcal{M}(a \to 1 + 2 + \cdots + N) \right|^2
\end{aligned}
} \quad (3.41)
$$

where j_a and λ_a are the spin and polarization of the initial state particle, and λ_i are the polarizations of the final state particles. In the previous expression we have averaged over initial state polarizations and summed over the final state polarizations. If we are interested in a polarized initial state we must drop the summation over λ_a and the $1/(2j_a + 1)$ average factor. If we are interested in polarized final states, then the summation over λ_1, λ_2, etc, must be dropped. Same consideration is valid for the cross section defined next

$$
\boxed{
\begin{aligned}
\sigma(a + b \to 1 + \cdots + N) &= \frac{1}{2\,\lambda^{1/2}(s, m_a^2, m_b^2)} \\
&\times \frac{1}{(2j_a + 1)(2j_b + 1)} \sum_{\lambda_a, \lambda_b, \lambda_1, \ldots, \lambda_N} \\
&\times \int dQ_N \left| \mathcal{M}(a + b \to 1 + \cdots + N) \right|^2
\end{aligned}
} \quad (3.42)
$$

Further Reading

A. Pich, Class Notes on Quantum Field Theory. http://eeemaster.uv.es/course/view.php?id=6
G. Kallen, *Elementary Particle Physics* (Addison-Wesley Publishing Company, Reading, 1964)
Particle Data Group
V. Barone, E. Predazzi, High-Energy Particle Diffraction

Chapter 4
Angular Distributions

Abstract In the previous chapter we presented the standard approach in defining kinematical variables and the phase-space. There is however, an alternative way of defining these variables and it is in terms of invariant masses and angles of pairs of particles in their center of mass reference system. This approach is very common for studying very rare decays (such as $B \to K^*\ell\ell$). Here we present the kinematics and phase space for one-to-three and one-to-four body decays.

4.1 Three Body Angular Distributions

Imagine a process like

$$\boxed{A(k) \; \to \; b(p) \; + \; c(p_1) \; + \; d(p_2)}, \tag{4.1}$$

where k, p, p_1 and p_2 are the momenta of the particles. We can choose one pair of particles (in this case we will choose c and d) and separate our decay (4.1) into two sequential decays. The first decay will be given by

where $q = p_1 + p_2$. Thus $A(k)$ decays into a *real* (on-shell) particle $b(p)$ and into a *virtual* set of particles $cd(q)$, which is characterised by the invariant mass (q^2) of the system of the two *real* (on-shell) particles $c(p_1)$ and $d(p_2)$. In the CM reference frame of A we have the following distribution of momenta

$$p^\mu\Big|_A = (p^0, \; 0, \; 0, \; -|\mathbf{q}|), \qquad q^\mu\Big|_A = (q^0, \; 0, \; 0, \; |\mathbf{q}|), \tag{4.2}$$

where we have supposed that b moves along the negative $\hat{z}$ axis and where p^0, q^0 and $|\mathbf{q}|$ are given by (see Chap. 3 for details)

© Springer International Publishing Switzerland 2016
V. Ilisie, *Concepts in Quantum Field Theory*,
UNITEXT for Physics, DOI 10.1007/978-3-319-22966-9_4

$$p^0 = \frac{1}{2m_A}\left(m_A^2 + m_b^2 - q^2\right), \tag{4.3}$$

$$q^0 = \frac{1}{2m_A}\left(m_A^2 - m_b^2 + q^2\right), \tag{4.4}$$

$$|\mathbf{q}| = \frac{1}{2m_A}\lambda^{1/2}\left(m_A^2, m_b^2, q^2\right). \tag{4.5}$$

It is straightforward to obtain the $p \cdot q$ invariant product

$$\boxed{p \cdot q = \frac{1}{2}\left(m_A^2 - m_b^2 - q^2\right).} \tag{4.6}$$

Going to the CM reference frame of the cd system we have the following configuration

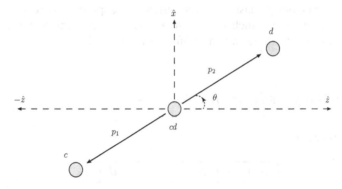

In this reference frame p and q are given by

$$p^\mu\Big|_{cd} = (p'^0,\ 0,\ 0,\ -|\mathbf{p}'|), \qquad q^\mu\Big|_{cd} = (\sqrt{q^2},\ 0,\ 0,\ 0). \tag{4.7}$$

Using the previously calculated $p \cdot q$ product and the expression $p^2 = m_b^2 = (p'^0)^2 - |\mathbf{p}'|^2$, we obtain

$$p'^0 = \frac{1}{2\sqrt{q^2}}\left(m_A^2 - m_b^2 - q^2\right), \tag{4.8}$$

$$|\mathbf{p}'| = \frac{1}{2\sqrt{q^2}}\lambda^{1/2}\left(m_A^2, m_b^2, q^2\right). \tag{4.9}$$

The four momenta p_1 and p_2 are given by

$$p_2^\mu\Big|_{cd} = (p_2^0,\ |\mathbf{p_2}|\sin\theta,\ 0,\ |\mathbf{p_2}|\cos\theta), \tag{4.10}$$

$$p_1^\mu\Big|_{cd} = (p_1^0,\ -|\mathbf{p_2}|\sin\theta,\ 0,\ -|\mathbf{p_2}|\cos\theta), \tag{4.11}$$

where

$$p_2^0 = \frac{1}{2\sqrt{q^2}} \left(q^2 + m_d^2 - m_c^2 \right),$$ (4.12)

$$p_1^0 = \frac{1}{2\sqrt{q^2}} \left(q^2 - m_d^2 + m_c^2 \right),$$ (4.13)

$$|\mathbf{p_2}| = \frac{1}{2\sqrt{q^2}} \lambda^{1/2} \left(q^2, m_c^2, m_d^2 \right).$$ (4.14)

It is straightforward to obtain the following invariant products:

$$
\boxed{
\begin{aligned}
p_1 \cdot q &= \frac{1}{2} (q^2 + m_c^2 - m_d^2) \\
p_2 \cdot q &= \frac{1}{2} (q^2 - m_c^2 + m_d^2) \\
p_1 \cdot p_2 &= \frac{1}{2} (q^2 - m_c^2 - m_d^2) \\
p \cdot p_1 &= p \cdot q - p \cdot p_2
\end{aligned}
}
$$ (4.15)

For the product $p \cdot p_2$ we simply have $p \cdot p_2 = p'^0 p_2^0 + |\mathbf{p'}| \, |\mathbf{p_2}| \cos \theta$ which explicitly reads

$$
\boxed{
\begin{aligned}
p \cdot p_2 &= \frac{1}{4q^2} \left(m_A^2 - m_b^2 - q^2 \right) \left(q^2 + m_d^2 - m_c^2 \right) \\
&\quad + \frac{1}{4q^2} \lambda^{1/2} \left(m_A^2, m_b^2, q^2 \right) \lambda^{1/2} \left(q^2, m_d^2, m_c^2 \right) \cos \theta
\end{aligned}
}
$$ (4.16)

When calculating the squared transition matrix for a given process, with the use of the invariant products given in the boxes one can express everything in terms of the masses m_i, the invariant squared mass of the cd system q^2, and $\cos \theta$. The only thing left is to properly define the corresponding phase space in terms of the same variables. The following formula will turn out to be extremely useful

$$\int d^4k \, \delta(k^2 - m^2) = \int \frac{d^3k}{2k^0},$$ (4.17)

where $k^0 > 0$. The generic expression for the three-body phase space is given by (see Chap. 3)

$$dQ_3 = S_n \frac{1}{(2\pi)^5} \frac{d^3p}{2p^0} \frac{d^3p_1}{2p_1^0} \frac{d^3p_2}{2p_2^0} \delta^{(4)}(k - p - p_1 - p_2)$$

$$= S_n \frac{1}{(2\pi)^5} \int d^4q \, \delta^{(4)}(q - p_1 - p_2) \frac{d^3p}{2p^0} \frac{d^3p_1}{2p_1^0} \frac{d^3p_2}{2p_2^0} \delta^{(4)}(k - p - q)$$

$$\equiv \frac{S_n}{2\pi} \times \mathcal{A} \times \mathcal{B}, \tag{4.18}$$

where we have defined $\mathcal{A}$ and $\mathcal{B}$ the following way

$$\mathcal{A} = \frac{1}{(2\pi)^2} \int d^4q \, \frac{d^3p}{2p^0} \delta^{(4)}(k - p - q), \tag{4.19}$$

$$\mathcal{B} = \frac{1}{(2\pi)^2} \frac{d^3p_1}{2p_1^0} \frac{d^3p_2}{2p_2^0} \delta^{(4)}(q - p_1 - p_2). \tag{4.20}$$

We will now manipulate these expressions to obtain the appropriate results in terms of the desired variables. For the first term we have

$$\mathcal{A} = \frac{1}{(2\pi)^2} \int dm_{12}^2 \int d^4q \, \frac{d^3p}{2p^0} \delta^{(4)}(k - p - q) \, \delta(q^2 - m_{12}^2)$$

$$= \frac{1}{(2\pi)^2} \int dm_{12}^2 \, \frac{d^3q}{2q^0} \frac{d^3p}{2p^0} \delta^{(4)}(k - p - q)$$

$$= \frac{1}{(2\pi)^2} \int dq^2 \, \frac{d^3q}{2q^0} \frac{d^3p}{2p^0} \delta^{(4)}(k - p - q)$$

$$= \frac{1}{8\pi m_A^2} \lambda^{1/2}\left(m_A^2, m_b^2, q^2\right) dq^2, \tag{4.21}$$

where, in order to get to the last line, we have used the expression (3.28) and we have integrated over the solid angle (without loss of generality). In order to simplify the notation we have also dropped the integral symbol in the last line. All we have left is to compute $\mathcal{B}$. Again, using (3.28) and integrating over ϕ (without loss of generality) we obtain

$$\mathcal{B} = \frac{1}{16\pi q^2} \lambda^{1/2}\left(q^2, m_c^2, m_d^2\right) d\cos\theta. \tag{4.22}$$

Finally, the expression of the differential three body phase space that we were looking for has the form

$$\boxed{dQ_3 = S_n \frac{1}{256\,\pi^3 \, m_A^2 \, q^2} \lambda^{1/2}\left(m_A^2, m_b^2, q^2\right) \lambda^{1/2}\left(q^2, m_c^2, m_d^2\right) dq^2 \, d\cos\theta}. \tag{4.23}$$

Fig. 4.1 Higgs-like scalar
particle ϕ decaying into a
pair of W gauge bosons

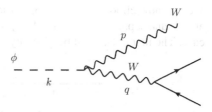

It is worth making the following comment. If the process we are studying is similar to the one shown in Fig. 4.1, then q^2 is the momentum of the virtual W boson. If however, $m_\phi > 2m_W$ then instead of a real (on-shell) and a virtual (off-shell) W boson we have two on-shell particles, thus $q^2 = m_W^2$. Therefore, the squared transition matrix will contain a squared propagator of the type

$$\frac{1}{\left(q^2 - m_W^2\right)^2 + m_W^2\, \Gamma_W^2} \tag{4.24}$$

which is regulated by a Breit-Wigner term, as a first order approximation (for more details see Chaps. 7 and 12). Working in the narrow width approximation we have

$$\lim_{\Gamma_W \to 0} \frac{1}{\left(q^2 - m_W^2\right)^2 + m_W^2\, \Gamma_W^2} = \frac{\pi}{m_W\, \Gamma_W}\, \delta\left(q^2 - m_W^2\right). \tag{4.25}$$

The resulting Dirac delta function can be simply reabsorbed into the definition of the phase space and we can perform the integration over q^2. Thus, our previous expression of dQ_3 would turn into

$$dQ_3 = \frac{1}{256\,\pi^3\, m_\phi^2\, m_W^2}\, \lambda^{1/2}\left(m_\phi^2, m_W^2, m_W^2\right) \lambda^{1/2}\left(m_W^2, m_c^2, m_d^2\right)\, d\cos\theta. \tag{4.26}$$

4.2 Four Body Angular Distributions

Imagine that in the previous process b is not a real particle, but represents a virtual set of two particles ($\tilde{c}\tilde{d}$) i.e., the process is given by

$$\boxed{A(k) \;\to\; \tilde{c}(\tilde{p}_1) \;+\; \tilde{d}(\tilde{p}_2) \;+\; c(p_1) \;+\; d(p_2)}. \tag{4.27}$$

We can now choose two pairs of particles (one given by c and d and another by $\tilde{c}$ and $\tilde{d}$) and separate our decay (4.27) into sequential decays, similar to the previous case. The first decay will be given by

$$\tilde{c}\tilde{d} \xleftarrow{\quad p \quad} A \xrightarrow{\quad q \quad} cd$$

where $q = p_1 + p_2$ and $p = \tilde{p}_1 + \tilde{p}_2$. Thus $A(k)$ decays into two *virtual* sets of particles $cd(q)$ and $\tilde{c}\tilde{d}(p)$ which are characterised by their invariant squared masses q^2 and p^2. Just as in the previous case the four momenta p and q in the CM reference frame of A are given by

$$p^\mu \Big|_A = (p^0,\ 0,\ 0,\ -|\mathbf{q}|), \qquad q^\mu \Big|_A = (q^0,\ 0,\ 0,\ |\mathbf{q}|), \qquad (4.28)$$

where, again, we have supposed the b moves along the negative $\hat{z}$ axis and where, this time

$$p^0 = \frac{1}{2m_A}\left(m_A^2 + p^2 - q^2\right), \qquad (4.29)$$

$$q^0 = \frac{1}{2m_A}\left(m_A^2 - p^2 + q^2\right), \qquad (4.30)$$

$$|\mathbf{q}| = \frac{1}{2m_A}\lambda^{1/2}\left(m_A^2, p^2, q^2\right). \qquad (4.31)$$

Thus, the invariant product $p \cdot q$ is simply

$$\boxed{p \cdot q = \frac{1}{2}\left(m_A^2 - p^2 - q^2\right).} \qquad (4.32)$$

The complete distribution of angles and momenta is schematically shown in Fig. 4.2. Thus, in the CM reference frame of the cd system, we have the following expressions for the four-momenta

$$p^\mu \Big|_{cd} = (p'^0,\ 0,\ 0,\ -|\mathbf{p}'|), \qquad q^\mu \Big|_{cd} = (\sqrt{q^2},\ 0,\ 0,\ 0). \qquad (4.33)$$

with

$$p'^0 = \frac{1}{2\sqrt{q^2}}\left(m_A^2 - p^2 - q^2\right), \qquad (4.34)$$

$$|\mathbf{p}'| = \frac{1}{2\sqrt{q^2}}\lambda^{1/2}\left(m_A^2, p^2, q^2\right). \qquad (4.35)$$

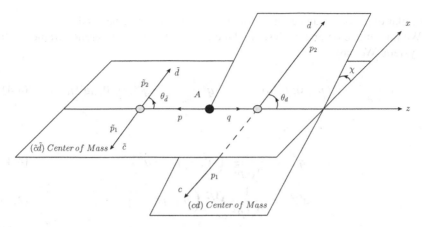

Fig. 4.2 Four body decay angular distribution

The four-momenta p_1 and p_2 simply read

$$p_2^{\mu}\Big|_{cd} = (p_2^0,\ |\mathbf{p_2}|\sin\theta_d\cos\chi,\ |\mathbf{p_2}|\sin\theta_d\sin\chi,\ |\mathbf{p_2}|\cos\theta_d), \qquad (4.36)$$

$$p_1^{\mu}\Big|_{cd} = (p_1^0,\ -|\mathbf{p_2}|\sin\theta_d\cos\chi,\ -|\mathbf{p_2}|\sin\theta_d\sin\chi,\ -|\mathbf{p_2}|\cos\theta_d), \qquad (4.37)$$

with p_2^0, p_1^0, and $|\mathbf{p_2}|$ given by the same expressions (4.12)–(4.14) from the previous section. One obviously obtains the same results for the scalar products (4.15) and

$$
\begin{aligned}
p \cdot p_2 = &\ \frac{1}{4q^2}\left(m_A^2 - p^2 - q^2\right)\left(q^2 + m_d^2 - m_c^2\right) \\
&+ \frac{1}{4q^2}\,\lambda^{1/2}\left(m_A^2, p^2, q^2\right)\lambda^{1/2}\left(q^2, m_d^2, m_c^2\right)\cos\theta_d
\end{aligned}
\qquad (4.38)
$$

Boosting[1] p_2^{μ} into the CM reference frame of A we obtain

$$
\begin{aligned}
p_2^{\mu}\Big|_{A} = &\ (\gamma\, p_2^0 + \gamma\, v\, |\mathbf{p_2}|\cos\theta_d,\ |\mathbf{p_2}|\sin\theta_d\cos\chi, \\
&\ |\mathbf{p_2}|\sin\theta_d\sin\chi,\ \gamma\, v\, p_2^0 + \gamma\,|\mathbf{p_2}|\cos\theta_d),
\end{aligned}
\qquad (4.39)
$$

with

$$v = \frac{|\mathbf{q}|}{q^0},\qquad\qquad \gamma = \frac{q^0}{\sqrt{q^2}}. \qquad (4.40)$$

---

[1] See Chap. 3 for the explicit expression for Lorentz boosts.

A similar expression can be found for p_1^μ, however it won't be needed.

We now move on and analyse the kinematics in the CM reference frame of the $\tilde{c}\tilde{d}$ system. We have

$$p^\mu \Big|_{\tilde{c}\tilde{d}} = (\sqrt{p^2}, 0, 0, 0), \qquad q^\mu \Big|_{\tilde{c}\tilde{d}} = (q'^0, 0, 0, |\mathbf{q}'|), \tag{4.41}$$

with

$$q'^0 = \frac{1}{2\sqrt{p^2}} \left(m_A^2 - p^2 - q^2\right), \tag{4.42}$$

$$|\mathbf{q}'|^2 = \frac{1}{2\sqrt{p^2}} \lambda^{1/2} \left(m_A^2, p^2, q^2\right). \tag{4.43}$$

For $\tilde{p}_1$ and $\tilde{p}_2$ we simply have the following expressions

$$\tilde{p}_2^\mu \Big|_{\tilde{c}\tilde{d}} = (\tilde{p}_2^0, |\tilde{\mathbf{p}}_2| \sin\theta_{\tilde{d}}, 0, |\tilde{\mathbf{p}}_2| \cos\theta_{\tilde{d}}), \tag{4.44}$$

$$\tilde{p}_1^\mu \Big|_{\tilde{c}\tilde{d}} = (\tilde{p}_1^0, -|\tilde{\mathbf{p}}_2| \sin\theta_{\tilde{d}}, 0, -|\tilde{\mathbf{p}}_2| \cos\theta_{\tilde{d}}). \tag{4.45}$$

The components of the four-momenta are again straightforward to obtain

$$\tilde{p}_2^0 = \frac{1}{2\sqrt{p^2}} \left(p^2 + m_{\tilde{d}}^2 - m_{\tilde{c}}^2\right), \tag{4.46}$$

$$\tilde{p}_1^0 = \frac{1}{2\sqrt{p^2}} \left(p^2 - m_{\tilde{d}}^2 + m_{\tilde{c}}^2\right), \tag{4.47}$$

$$|\tilde{\mathbf{p}}_2| = \frac{1}{2\sqrt{p^2}} \lambda^{1/2} \left(p^2, m_{\tilde{c}}^2, m_{\tilde{d}}^2\right). \tag{4.48}$$

We can now calculate, as in the cd system, similar invariant scalar products

$$\boxed{\begin{aligned} \tilde{p}_1 \cdot p &= \frac{1}{2} (p^2 + m_{\tilde{c}}^2 - m_{\tilde{d}}^2) \\ \tilde{p}_2 \cdot p &= \frac{1}{2} (p^2 - m_{\tilde{c}}^2 + m_{\tilde{d}}^2) \\ \tilde{p}_1 \cdot \tilde{p}_2 &= \frac{1}{2} (p^2 - m_{\tilde{c}}^2 - m_{\tilde{d}}^2) \\ q \cdot \tilde{p}_1 &= p \cdot q - q \cdot \tilde{p}_2 \end{aligned}} \tag{4.49}$$

with

$$
\boxed{
\begin{aligned}
q \cdot \tilde{p}_2 &= \frac{1}{4p^2}(m_A^2 - p^2 - q^2)(p^2 + m_{\tilde{d}}^2 - m_{\tilde{c}}^2) \\
&\quad - \frac{1}{4p^2}\lambda^{1/2}(m_A^2, p^2, q^2)\,\lambda^{1/2}(p^2, m_{\tilde{d}}^2, m_{\tilde{c}}^2)\cos\theta_{\tilde{d}}
\end{aligned}
}
\tag{4.50}
$$

Using momentum conservation we can also calculate products of the type $p_i \cdot \tilde{p}_j$. We find

$$
\boxed{
\begin{aligned}
p_1 \cdot \tilde{p}_2 &= q \cdot \tilde{p}_2 - p_2 \cdot \tilde{p}_2 \\
p_2 \cdot \tilde{p}_1 &= p \cdot p_2 - p_2 \cdot \tilde{p}_2 \\
p_1 \cdot \tilde{p}_1 &= p \cdot p_1 - q \cdot \tilde{p}_2 + p_2 \cdot \tilde{p}_2
\end{aligned}
}
\tag{4.51}
$$

The remaining product we need to calculate is $p_2 \cdot \tilde{p}_2$. Boosting $\tilde{p}_2^\mu$ into the CM frame of A we obtain

$$
\tilde{p}_2^\mu \Big|_A = (\tilde{\gamma}\,\tilde{p}_2^0 - \tilde{\gamma}\,\tilde{v}\,|\tilde{\mathbf{p}}_2|\cos\theta_{\tilde{d}},\ |\tilde{\mathbf{p}}_2|\sin\theta_{\tilde{d}},\ 0,\ -\tilde{\gamma}\,\tilde{v}\,\tilde{p}_2^0 + \tilde{\gamma}\,|\tilde{\mathbf{p}}_2|\cos\theta_{\tilde{d}}),
\tag{4.52}
$$

with

$$
\tilde{\gamma} = \frac{p^0}{\sqrt{p^2}},\qquad \tilde{v} = \frac{|\mathbf{q}|}{p^0}.
\tag{4.53}
$$

Using the expressions (4.39) and (4.52) we can now evaluate the $p_2 \cdot \tilde{p}_2$ invariant product in the CM reference system of A

$$
\boxed{
\begin{aligned}
p_2 \cdot \tilde{p}_2 &= (\gamma\,p_2^0 + \gamma\,v\,|\mathbf{p}_2|\cos\theta_d)(\tilde{\gamma}\,\tilde{p}_2^0 - \tilde{\gamma}\,\tilde{v}\,|\tilde{\mathbf{p}}_2|\cos\theta_{\tilde{d}}) \\
&\quad - |\mathbf{p}_2|\,|\tilde{\mathbf{p}}_2|\sin\theta_{\tilde{d}}\sin\theta_d\cos\chi \\
&\quad + (\gamma\,v\,p_2^0 + \gamma\,|\mathbf{p}_2|\cos\theta_d)(\tilde{\gamma}\,\tilde{v}\,\tilde{p}_2^0 - \tilde{\gamma}\,|\tilde{\mathbf{p}}_2|\cos\theta_{\tilde{d}})
\end{aligned}
}
\tag{4.54}
$$

One can now express all the terms from the squared transition matrix (by using the expressions in the boxes) in terms of the squared invariant masses, p^2, q^2, the masses m_i and the three angles θ_d, $\theta_{\tilde{d}}$ and χ. Furthermore, one can use the **helicity amplitudes** approach in order to logically and systematically group all the terms from the transition matrix and obtain the so called angular coefficients.[2]

---

[2]For more details check Furhter Reading.

Last we will calculate the needed phase space. According to the definition given in the previous chapter, the generic expression for the four-body phase space is given by

$$
\begin{aligned}
dQ_4 &= S_n \frac{1}{(2\pi)^8} \frac{d^3 p_1}{2p_1^0} \frac{d^3 p_2}{2p_2^0} \frac{d^3 \tilde{p}_1}{2\tilde{p}_1^0} \frac{d^3 \tilde{p}_2}{2\tilde{p}_2^0} \delta^{(4)}(k - p_1 - p_2 - \tilde{p}_1 - \tilde{p}_2) \\
&= S_n \frac{1}{(2\pi)^8} \int d^4 p \int d^4 q \, \delta^{(4)}(q - p_1 - p_2) \, \delta^{(4)}(p - \tilde{p}_1 - \tilde{p}_2) \\
&\qquad\qquad\qquad \times \frac{d^3 p_1}{2p_1^0} \frac{d^3 p_2}{2p_2^0} \frac{d^3 \tilde{p}_1}{2\tilde{p}_1^0} \frac{d^3 \tilde{p}_2}{2\tilde{p}_2^0} \delta^{(4)}(k - p - q) \\
&\equiv \frac{S_n}{(2\pi)^2} \times \mathcal{A} \times \mathcal{B} \times \mathcal{C},
\end{aligned}
\tag{4.55}
$$

where we have defined $\mathcal{A}$, $\mathcal{B}$ and $\mathcal{C}$ the following way

$$
\mathcal{A} = \frac{1}{(2\pi)^2} \int d^4 p \int d^4 q \, \delta^{(4)}(k - p - q),
\tag{4.56}
$$

$$
\mathcal{B} = \frac{1}{(2\pi)^2} \frac{d^3 p_1}{2p_1^0} \frac{d^3 p_2}{2p_2^0} \delta^{(4)}(q - p_1 - p_2),
\tag{4.57}
$$

$$
\mathcal{C} = \frac{1}{(2\pi)^2} \frac{d^3 \tilde{p}_1}{2\tilde{p}_1^0} \frac{d^3 \tilde{p}_2}{2\tilde{p}_2^0} \delta^{(4)}(p - \tilde{p}_1 - \tilde{p}_2).
\tag{4.58}
$$

Using the same techniques as in the previous section we can manipulate $\mathcal{A}$

$$
\begin{aligned}
\mathcal{A} &= \frac{1}{(2\pi)^2} \int dm_{12}^2 \int d\tilde{m}_{12}^2 \int d^4 p \int d^4 q \, \delta^{(4)}(k - p - q) \\
&\qquad\qquad\qquad\qquad\qquad \times \delta(q^2 - m_{12}^2) \, \delta(p^2 - \tilde{m}_{12}^2) \\
&= \frac{1}{(2\pi)^2} \int dm_{12}^2 \int d\tilde{m}_{12}^2 \int \frac{d^3 p}{2p^0} \int \frac{d^3 q}{2q^0} \, \delta^{(4)}(k - p - q) \\
&= \frac{1}{(2\pi)^2} \int dq^2 \int dp^2 \int \frac{d^3 p}{2p^0} \int \frac{d^3 q}{2q^0} \, \delta^{(4)}(k - p - q) \\
&= \frac{1}{8\pi m_A^2} \lambda^{1/2}(m_A^2, q^2, p^2) \, dq^2 \, dp^2.
\end{aligned}
\tag{4.59}
$$

Again, we have used (3.28), integrated over the solid angle (without loss of generality) and dropped the integral symbols to simplify notation. For $\mathcal{B}$ we can still integrate over the azimuthal angle without loosing generality, and we obtain

$$
\mathcal{B} = \frac{1}{16\pi q^2} \lambda^{1/2}\left(q^2, m_d^2, m_c^2\right) d\cos\theta_d.
\tag{4.60}
$$

Finally, for $\mathcal{C}$ we must keep both angles

$$\mathcal{C} = \frac{1}{32\pi^2 p^2} \lambda^{1/2} \left(p^2, m_{\bar{d}}^2, m_{\bar{c}}^2 \right) d\cos\theta_{\bar{d}}\, d\chi. \tag{4.61}$$

Thus, our final expression of the four-body phase space in terms of angular variables is given by

$$
\begin{aligned}
dQ_4 &= S_n \frac{1}{(128\pi^3)^2\, m_A^2\, p^2\, q^2} \lambda^{1/2}(q^2, m_d^2, m_c^2)\, \lambda^{1/2}(p^2, m_{\bar{d}}^2, m_{\bar{c}}^2) \\
&\qquad\qquad \times\, dq^2\, dp^2\, d\cos\theta_d\, d\cos\theta_{\bar{d}}\, d\chi
\end{aligned}
\tag{4.62}
$$

If we are facing with a similar case as in the previous section, p^2 or q^2 or both are the momenta of some virtual particles that can reach the on-shell region, and we are working in the narrow width approximation, we can simply reabsorb the Dirac delta function(s) into dQ_4 and integrate (just as in the previous section) with no further complication.

Further Reading

N. Cabibbo, A. Maksymowicz, Angular correlations in Ke-4 decays and determination of low-energy pi-pi phase shifts. Phys. Rev. **137**, B438 (1965)

A. Faessler, T. Gutsche, M.A. Ivanov, J.G. Korner, V.E. Lyubovitskij, The Exclusive rare decays $B \to K(K^*)\, \bar{\ell}\ell$ and $B_c \to D(D^*)\, \bar{\ell}\ell$ in a relativistic quark model. Eur. Phys. J. direct C **4**, 18 (2002). arXiv:hep-ph/0205287

G. Kallen, *Elementary Particle Physics* (Publishing Company, Addison-Wesley, 1964)

W. Altmannshofer, P. Ball, A. Bharucha, A.J. Buras, D.M. Straub and M. Wick, Symmetries and asymmetries of $B \to K^*\mu^+\mu^-$ decays in the standard model and beyond. JHEP **0901**, 019 (2009). http://arxiv.org/abs/pdf/0811.1214.pdf

Chapter 5
Dirac Algebra

Abstract In this chapter we present the basics of Dirac spinor algebra needed for calculations involving fermions. We introduce the commuting and anti-commuting relations among the various Dirac matrices and we present the basics of calculating spinor traces. The tools given here can be used to further calculate lengthy and more complicated traces involving Dirac matrices. The transformation of spinors under Lorentz transformations is also presented consistently together with the bilinear covariants. Finally a short comparison between QED and QCD is given for a simple process.

5.1 Dirac Matrices

We shall start this chapter by introducing the well known Dirac equation. It reads

$$\boxed{\left(i\,\partial\!\!\!/ - m\right)\psi(x) = 0},\qquad (5.1)$$

where we have used the Feynman slashed notation $\partial\!\!\!/ \equiv \gamma^\mu \partial_\mu$. The γ^μ Dirac matrices obey the Clifford algebra

$$\frac{1}{2}\{\gamma^\mu,\gamma^\nu\} \equiv \frac{1}{2}\left(\gamma^\mu\gamma^\nu + \gamma^\nu\gamma^\mu\right) = g^{\mu\nu}\,I_4,\qquad \gamma^{\mu\dagger} = \gamma^0\gamma^\mu\gamma^0.\qquad (5.2)$$

Thus, it is straightforward to deduce

$$\left(\gamma^0\right)^2 = -\left(\gamma^i\right)^2 = I_4,\qquad \gamma^{0\dagger} = \gamma^0,\qquad \gamma^{i\dagger} = -\gamma^i,\qquad (5.3)$$

where I_4 is the identity matrix in 4 dimensions (that we shall not write down explicitly unless necessary). Using the previously introduced γ^μ matrices one can further construct two others. The first one

$$\gamma_5 \equiv i\,\gamma^0\gamma^1\gamma^2\gamma^3 = -\frac{i}{4}\,\epsilon_{\mu\nu\alpha\beta}\,\gamma^\mu\gamma^\nu\gamma^\alpha\gamma^\beta,\qquad (5.4)$$

© Springer International Publishing Switzerland 2016
V. Ilisie, *Concepts in Quantum Field Theory*,
UNITEXT for Physics, DOI 10.1007/978-3-319-22966-9_5

with $\epsilon^{0123} = -\epsilon_{0123} = 1$, and the second one

$$\sigma^{\mu\nu} \equiv \frac{i}{2}[\gamma^\mu, \gamma^\nu] \equiv \frac{i}{2}\left(\gamma^\mu\gamma^\nu - \gamma^\nu\gamma^\mu\right). \tag{5.5}$$

It is easy to check that

$$\gamma_5\,\sigma^{\mu\nu} = \frac{i}{2}\,\epsilon^{\mu\nu\alpha\beta}\,\sigma_{\alpha\beta}, \qquad \left(\gamma_5\right)^2 = I_4. \tag{5.6}$$

In the previous expression we have defined $\sigma_{\alpha\beta} \equiv g_{\rho\alpha}\,g_{\delta\beta}\,\sigma^{\rho\delta}$. Using the Clifford algebra one can easily prove

$$\gamma_5^\dagger = \gamma_5, \qquad \sigma^{\mu\nu\dagger} = \gamma^0\,\sigma^{\mu\nu}\,\gamma^0. \tag{5.7}$$

and

$$\{\gamma_5, \gamma^\mu\} = 0, \qquad [\gamma_5, \sigma^{\mu\nu}] = 0. \tag{5.8}$$

Thus, using the Clifford algebra and the previously introduced $\sigma^{\mu\nu}$ matrix, one can write $\gamma^\mu\gamma^\nu$ as

$$\boxed{\gamma^\mu\gamma^\nu = 2g^{\mu\nu} - \gamma^\nu\gamma^\mu = g^{\mu\nu} - i\,\sigma^{\mu\nu}}. \tag{5.9}$$

With all the previously introduced concepts one can calculate any tensor contraction i.e.,

$$\gamma^\mu\,\gamma_\mu = 4\,I_4\,, \qquad \gamma^\mu\,\gamma^\nu\,\gamma_\mu = -2\gamma^\nu\,, \qquad \gamma^\mu\,\slashed{p}\,\gamma_\mu = -2\slashed{p}, \tag{5.10}$$

etc., or products of the type

$$\slashed{p}\slashed{q} = p\cdot q - i\,\sigma^{\mu\nu}\,p_\mu q_\nu, \tag{5.11}$$

or tensor identities like

$$\gamma^\mu\,\gamma^\nu\,\gamma^\rho + \gamma^\rho\,\gamma^\nu\,\gamma^\mu = 2\left(g^{\mu\nu}\,\gamma^\rho - g^{\mu\rho}\,\gamma^\nu + g^{\nu\rho}\,\gamma^\mu\right). \tag{5.12}$$

The following results involving the Levi-Civita tensor density[1] will also turn out to be useful:

$$\epsilon^{\alpha\beta\mu\nu}\,\epsilon_{\alpha\beta\sigma\rho} = 2\left(\delta^\mu_\rho\,\delta^\nu_\sigma - \delta^\mu_\sigma\,\delta^\nu_\rho\right), \tag{5.13}$$

$$\epsilon^{\alpha\beta\rho\mu}\,\epsilon_{\alpha\beta\rho\nu} = -6\,\delta^\mu_\nu, \tag{5.14}$$

$$\epsilon^{\alpha\beta\mu\nu}\,\epsilon_{\alpha\beta\mu\nu} = -24. \tag{5.15}$$

---

[1] See Chap. 1 for the definition of tensor density.

Some authors use g_ν^μ instead of the usual Kronecker delta tensor δ_ν^μ. This should only be interpreted as a mnemotechnical rule for raising (or lowering) tensor indices i.e., $g_\nu^\mu \to g^{\mu\nu}$. Thus, it is used to remind that when raising an index of the Kronecker delta tensor it *transforms* into the metric tensor i.e., $g^{\mu\alpha}\,\delta_\alpha^\nu = g^{\mu\nu}$.

5.2 Dirac Traces

When computing transition matrix elements involving fermions, one usually needs to calculate traces of products of Dirac matrices. One could find useful the following generic trace properties

$$\mathrm{Tr}\{A\} = \mathrm{Tr}\{A^T\}, \qquad \mathrm{Tr}\{A + B\} = \mathrm{Tr}\{A\} + \mathrm{Tr}\{B\}, \qquad (5.16)$$

where T stands for transposed, and

$$\begin{aligned}\mathrm{Tr}\{A_1 \ldots A_N\} &= \mathrm{Tr}\{A_N\,A_1 \ldots A_{N-1}\} \\ &= \mathrm{Tr}\{A_{N-1}\,A_N\,A_1 \ldots A_{N-2}\}, \qquad (5.17)\end{aligned}$$

etc., which is called the cyclic property. Returning to the Dirac matrices, using the previous results one can calculate any trace of Dirac matrices. Some interesting simple results are presented next.

$$\mathrm{Tr}\{\gamma_5\} = \mathrm{Tr}\{\gamma^\mu\} = \mathrm{Tr}\{\sigma^{\mu\nu}\} = 0. \qquad (5.18)$$

For an odd number of Dirac matrices one finds ($k \in \mathbb{N}$)

$$\mathrm{Tr}\{\gamma^{\mu_1} \ldots \gamma^{\mu_{2k+1}}\} = 0, \qquad (5.19)$$

therefore, the following equality also holds

$$\mathrm{Tr}\{\gamma_5\,\gamma^{\mu_1} \ldots \gamma^{\mu_{2k+1}}\} = 0. \qquad (5.20)$$

Other interesting results are

$$\mathrm{Tr}\{\gamma^\mu\,\gamma^\nu\} = 4\,g^{\mu\nu}, \qquad (5.21)$$

$$\mathrm{Tr}\{\gamma^\mu\,\gamma_\mu\} = 16, \qquad (5.22)$$

$$\mathrm{Tr}\{\gamma_5\,\gamma^\mu\,\gamma^\nu\} = 0, \qquad (5.23)$$

$$\mathrm{Tr}\{\gamma^\mu\,\gamma^\nu\,\gamma^\rho\,\gamma^\sigma\} = 4\left(g^{\mu\nu}\,g^{\sigma\rho} - g^{\mu\rho}\,g^{\nu\sigma} + g^{\mu\sigma}\,g^{\nu\rho}\right), \qquad (5.24)$$

$$\mathrm{Tr}\{\gamma_5\,\gamma^\mu\,\gamma^\nu\,\gamma^\rho\,\gamma^\sigma\} = -4\,i\,\epsilon^{\mu\nu\rho\sigma}. \qquad (5.25)$$

With the tools given here, one can calculate more lengthy and complicated traces involving Dirac matrices with no further complication.

5.3 Spinors and Lorentz Transformations

In this section we shall study the transformation of a spinor field $\psi(x)$ under a Lorentz transformation

$$x^\mu \;\rightarrow\; x'^\mu \;=\; \Lambda^\mu{}_\nu x^\nu . \tag{5.26}$$

Under the previous transformation a spinor field should transform as

$$\boxed{\psi(x) \;\rightarrow\; \psi'(x') \;=\; \psi'(\Lambda x) \;\equiv\; S(\Lambda)\,\psi(x)}, \tag{5.27}$$

where we have supposed there exists a linear operator of the form $S(\Lambda)$ that implements the Lorentz transformation Λ on the spinor field $\psi(x)$. Let's further analyse the properties that this operator should have:

$$\psi'(x') \;=\; S(\Lambda)\,\psi(x) \;=\; S(\Lambda)\,\psi(\Lambda^{-1} x) \;=\; S(\Lambda)\,S(\Lambda^{-1})\,\psi'(x'). \tag{5.28}$$

Therefore we have found that $S(\Lambda^{-1}) = S^{-1}(\Lambda)$. Let's move on and see what are the implications of this operator on the Dirac matrices. One of the basic principles of Relativity (Special or General) is that all equations of motion must have the same *form* in all reference frames. Thus the Dirac equation

$$\bigl(i\,\gamma^\mu\,\partial_\mu - m\bigr)\psi(x) \;=\; 0, \tag{5.29}$$

in terms of the primed variables must be written as

$$\bigl(i\,\gamma^\nu\,\partial'_\nu - m\bigr)\psi'(x') \;=\; 0. \tag{5.30}$$

where $\partial'_\nu \equiv \partial/\partial x'^\nu$. We find the following

$$\begin{aligned}
0 &= \Bigl(i\,\gamma^\mu\,\partial_\mu - m\Bigr)\psi(x) \\
&= \Bigl(i\,\gamma^\mu\,\partial_\mu - m\Bigr) S^{-1}(\Lambda)\,\psi'(x') \\
&= \Bigl(i\,S(\Lambda)\,\gamma^\mu\,S^{-1}(\Lambda)\,\partial_\mu - m\Bigr)\psi'(x') \\
&= \Bigl(i\,S(\Lambda)\,\gamma^\mu\,S^{-1}(\Lambda)\,\Lambda^\nu{}_\mu\,\partial'_\nu - m\Bigr)\psi'(x') \\
&= \Bigl(i\,\gamma^\nu\,\partial'_\nu - m\Bigr)\psi'(x').
\end{aligned} \tag{5.31}$$

Thus, under a Lorentz transformation (5.26) the Dirac gamma matrices are related through

$$\boxed{\gamma^\nu = S(\Lambda) \left(\Lambda^\nu{}_\mu \gamma^\mu \right) S^{-1}(\Lambda)}. \tag{5.32}$$

Let's now try to find and explicit expression for $S(\Lambda)$ for a proper orthochronous Lorentz transformation (1.124)

$$\Lambda^\mu{}_\nu = \delta^\mu_\nu + \Delta\omega^\mu{}_\nu + \mathcal{O}(\Delta\omega^2). \tag{5.33}$$

We write down the ansatz

$$S(\Lambda) = I_4 + b_{\mu\nu} \Delta\omega^{\mu\nu} + \mathcal{O}(\Delta\omega^2), \tag{5.34}$$

$$S^{-1}(\Lambda) = I_4 - b_{\mu\nu} \Delta\omega^{\mu\nu} + \mathcal{O}(\Delta\omega^2). \tag{5.35}$$

(with $\Delta\omega^{\mu\nu} \equiv g^{\alpha\nu} \Delta\omega^\mu{}_\alpha$) where $b_{\mu\nu}$ must be 4×4 antisymmetric matrices (in the μ, ν indices due to the fact that $\Delta\omega^{\mu\nu}$ is also antisymmetric). Thus, the relation (5.32) takes the form

$$\gamma^\nu = \left(I_4 + b_{\sigma\rho} \Delta\omega^{\sigma\rho} \right) \left(\gamma^\nu + \Delta\omega^\nu{}_\mu \gamma^\mu \right) \left(I_4 - b_{\alpha\beta} \Delta\omega^{\alpha\beta} \right) + \cdots \tag{5.36}$$

Keeping only terms up to $\mathcal{O}(\Delta\omega)$ we obtain

$$\Delta\omega^\nu{}_\mu \gamma^\mu = \gamma^\nu b_{\alpha\beta} \Delta\omega^{\alpha\beta} - b_{\sigma\rho} \Delta\omega^{\sigma\rho} \gamma^\nu$$

$$= \Delta\omega^{\alpha\beta} [\gamma^\nu, b_{\alpha\beta}]. \tag{5.37}$$

On the other hand we can write $\Delta\omega^\nu{}_\mu \gamma^\mu$ as

$$\Delta\omega^\nu{}_\mu \gamma^\mu = \frac{1}{2} \Delta\omega^{\alpha\beta} \left(\gamma_\beta \delta^\nu_\alpha - \gamma_\alpha \delta^\nu_\beta \right). \tag{5.38}$$

Comparing the last two equations (for arbitrary $\Delta\omega^{\alpha\beta}$) we find

$$\frac{1}{2} \left(\gamma_\beta \delta^\nu_\alpha - \gamma_\alpha \delta^\nu_\beta \right) = [\gamma^\nu, b_{\alpha\beta}]. \tag{5.39}$$

It is easy to check that this relation is satisfied for

$$b_{\alpha\beta} = -\frac{i}{4} \sigma_{\alpha\beta}. \tag{5.40}$$

Thus, the operator that implements an infinitesimal proper orthochronus Lorentz transformation on the spinor field is given by

$$
S(\Lambda) \; = \; I_4 - \frac{i}{4}\sigma_{\mu\nu}\,\Delta\omega^{\mu\nu} + \ldots ,
\tag{5.41}
$$

with its inverse given by

$$
S^{-1}(\Lambda) \; = \; I_4 + \frac{i}{4}\sigma_{\mu\nu}\,\Delta\omega^{\mu\nu} + \cdots \; = \; \gamma^0\, S^\dagger\, \gamma^0 .
\tag{5.42}
$$

For a parity transformation (1.127) we simply find

$$
S\big(\Lambda(P)\big) \; = \; S^{-1}\big(\Lambda(P)\big) \; = \; S^\dagger\big(\Lambda(P)\big) \; = \; \gamma^0 .
\tag{5.43}
$$

Defining $\overline{\psi}(x) \equiv \psi^\dagger(x)\,\gamma^0$, we obtain that under a Lorentz transformation this spinor transforms as

$$
\overline{\psi}'(x') \; = \; \overline{\psi}(x)\, S^{-1}(\Lambda) .
\tag{5.44}
$$

Restricting our transformations to proper orthochronous Lorentz boosts and parity we can construct the following bilinear covariants

$$
\overline{\psi}'(x')\,\psi'(x') = \overline{\psi}(x)\,\psi(x) \;\rightarrow\; \text{scalar},
\tag{5.45}
$$

$$
\overline{\psi}'(x')\,\gamma_5\,\psi'(x') = \det(\Lambda)\,\overline{\psi}(x)\,\gamma_5\,\psi(x) \;\rightarrow\; \text{scalar density},
\tag{5.46}
$$

$$
\overline{\psi}'(x')\,\gamma^\mu\,\psi'(x') = \Lambda^\mu{}_\nu\,\overline{\psi}(x)\,\gamma^\nu\,\psi(x) \;\rightarrow\; \text{vector},
\tag{5.47}
$$

$$
\overline{\psi}'(x')\,\gamma_5\,\gamma^\mu\,\psi'(x') = \det(\Lambda)\,\Lambda^\mu{}_\nu\,\overline{\psi}(x)\,\gamma_5\,\gamma^\nu\,\psi(x) \;\rightarrow\; \text{vector density},
\tag{5.48}
$$

$$
\overline{\psi}'(x')\,\sigma^{\mu\nu}\,\psi'(x') = \Lambda^\mu{}_\alpha\,\Lambda^\nu{}_\beta\,\overline{\psi}(x)\,\sigma^{\alpha\beta}\,\psi(x) \;\rightarrow\; \text{tensor}.
\tag{5.49}
$$

5.4 Quantum Electrodynamics

We are now ready to move on and apply all the spinor techniques learned in the previous sections to the calculation of scattering processes. We will turn our attention to the QED Lagrangian

$$
\mathcal{L}_{QED} = -\frac{1}{4}\,F_{\mu\nu}F^{\mu\nu} - \frac{1}{2\xi}\Big(\partial_\mu A^\mu\Big)^2 + i\,\overline{\psi}\,\gamma^\mu\,\partial_\mu\,\psi
$$
$$
- m\,\overline{\psi}\,\psi - e\,Q\,A_\mu\,\overline{\psi}\,\gamma^\mu\,\psi
\tag{5.50}
$$

As it is already well known, the QED interaction Lagrangian is obtained by imposing local gauge invariance on the Dirac field. Using QED as an inspirational toy-model is how the Standard Model was finally born. If one extends the local gauge invariance principle from the $U(1)$ group (that corresponds to QED) to the $SU(3)_C \otimes SU(2)_L \otimes U(1)_Y$ group and introduces the adequate fermion families, one obtains the the Standard Model Lagrangian.[2]

Returning to our simple QED Lagrangian, after quantization[3] (canonical, path integral, etc.) one can define the Feynman rules of the model. In our case they are given by

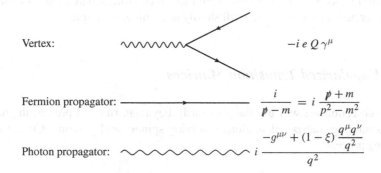

Vertex: $\qquad\qquad\qquad\qquad\qquad\qquad -i\,e\,Q\,\gamma^\mu$

Fermion propagator: $\qquad\qquad\qquad\qquad \dfrac{i}{\not{p}-m} = i\,\dfrac{\not{p}+m}{p^2-m^2}$

Photon propagator: $\qquad\qquad\qquad i\,\dfrac{-g^{\mu\nu}+(1-\xi)\dfrac{q^\mu q^\nu}{q^2}}{q^2}$

for the vertex and propagators, and

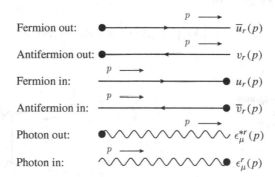

Fermion out: $\qquad\qquad p \longrightarrow \qquad \bar{u}_r(p)$

Antifermion out: $\qquad\qquad p \longrightarrow \qquad v_r(p)$

Fermion in: $\qquad p \longrightarrow \qquad\qquad u_r(p)$

Antifermion in: $\qquad p \longrightarrow \qquad\qquad \bar{v}_r(p)$

Photon out: $\qquad\qquad\qquad p \longrightarrow \qquad \epsilon_\mu^{*r}(p)$

Photon in: $\qquad\qquad p \longrightarrow \qquad\qquad \epsilon_\mu^{r}(p)$

for the external fields (the index r stands for the corresponding polarization). The Dirac spinors introduced previously satisfy the equations

---

[2]For a nice review of how this is done read A. Pich, *The Standard Model of Electroweak Interactions*, http://arxiv.org/pdf/1201.0537v1.pdf and A. Pich, *Quantum Chromodynamics*, http://arxiv.org/pdf/hep-ph/9505231.pdf.

[3]For more details one should consult the last four references at the end of this chapter.

$$\left(\not p - m\right)u_r(p) = 0, \qquad\qquad \bar{u}_r(p)\left(\not p - m\right) = 0, \qquad (5.51)$$

$$\left(\not p + m\right)v_r(p) = 0, \qquad\qquad \bar{v}_r(p)\left(\not p + m\right) = 0. \qquad (5.52)$$

Summing over the polarization of the spinors one finds the following relations

$$\boxed{\sum_{r=1}^{2} u_r(p)\,\bar{u}_r(p) = \left(\not p + m\right), \qquad \sum_{r=1}^{2} v_r(p)\,\bar{v}_r(p) = \left(\not p - m\right).} \qquad (5.53)$$

These two results will turn out to be useful for calculating squared matrix elements for unpolarized fermions, as we shall shortly see with an example.

5.4.1 Unpolarized Transition Matrices

In this subsection we will use the previously Feynman rules to present the basics of the scattering matrix calculations involving spinors and photons. Consider the following process

$$e^+(p_1)\,e^-(p_2) \;\rightarrow\; \gamma(k_1)\,\gamma(k_2). \qquad (5.54)$$

The corresponding Feynman diagrams are shown in Fig 5.1. The total squared transition matrix will be given by

$$|\mathcal{M}|^2 = |\mathcal{M}_1|^2 + |\mathcal{M}_2|^2 + 2\,\mathrm{Re}\,\mathcal{M}_1^\dagger\mathcal{M}_2. \qquad (5.55)$$

Fig. 5.1 Feynman diagrams for electron-positron annihilation to a pair of photons

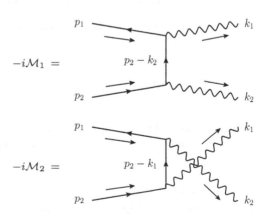

Here we shall only focus our attention on diagram (1). The rest of the calculation is left for the reader as an exercise. The transition matrix for the first diagram is given by

$$
\mathcal{M}_1 = e^2\, Q^2\, \epsilon_\mu^{r*}(k_1)\, \epsilon_\nu^{s*}(k_2)\, \frac{1}{(p_2 - k_2)^2 - m^2}\, \bar{v}_{r_1}(p_1)
$$

$$
\times\, \gamma^\mu\, (\not{p}_2 - \not{k}_2 + m)\, \gamma^\nu\, u_{r_2}(p_2). \tag{5.56}
$$

In the following we will explicitly calculate the hermitian conjugate of the spinor part of this transition matrix:

$$
\left(\bar{v}_{r_1}\, \gamma^\mu\, (\not{p}_2 - \not{k}_2 + m)\, \gamma^\nu\, u_{r_2} \right)^\dagger =
$$

$$
= \left(v_{r_1}^\dagger\, \gamma^0\, \gamma^\mu\, (\not{p}_2 - \not{k}_2 + m)\, \gamma^\nu\, u_{r_2} \right)^\dagger
$$

$$
= u_{r_2}^\dagger\, \gamma^{\nu\dagger}\, (\not{p}_2^\dagger - \not{k}_2^\dagger + m)\, \gamma^{\mu\dagger}\, \gamma^{0\dagger}\, v_{r_1}
$$

$$
= u_{r_2}^\dagger\, \gamma^0\, \gamma^\nu\, \gamma^0\, (\gamma^0\, \not{p}_2\, \gamma^0 - \gamma^0\, \not{k}_2\, \gamma^0 + m)\, \gamma^0\, \gamma^\mu\, \gamma^0\, \gamma^0\, v_{r_1}
$$

$$
= \bar{u}_{r_2}\, \gamma^\nu\, (\not{p}_2 - \not{k}_2 + m)\, \gamma^\mu\, v_{r_1}. \tag{5.57}
$$

Thus, the hermitian conjugate of $\mathcal{M}_1$ is simply given by

$$
\mathcal{M}_1^\dagger = e^2\, Q^2\, \epsilon_{\mu'}^{r}(k_1)\, \epsilon_{\nu'}^{s}(k_2)\, \frac{1}{(p_2 - k_2)^2 - m^2}\, \bar{u}_{r_2}(p_2)
$$

$$
\times\, \gamma^{\nu'}\, (\not{p}_2 - \not{k}_2 + m)\, \gamma^{\mu'}\, v_{r_1}(p_1). \tag{5.58}
$$

The squared transition matrix for the first diagram then reads

$$
|\mathcal{M}_1|^2 = e^4\, Q^4\, \epsilon_\mu^{r*}(k_1)\, \epsilon_{\mu'}^{r}(k_1)\, \epsilon_\nu^{s*}(k_2)\, \epsilon_{\nu'}^{s}(k_2)\, \frac{1}{[(p_2 - k_2)^2 - m^2]^2}
$$

$$
\times\, \bar{u}_{r_2}\, \gamma^{\nu'}\, (\not{p}_2 - \not{k}_2 + m)\, \gamma^{\mu'}\, v_{r_1}\, \bar{v}_{r_1}\, \gamma^\mu\, (\not{p}_2 - \not{k}_2 + m)\, \gamma^\nu\, u_{r_2}. \tag{5.59}
$$

Introducing explicitly the matrix indices (the upper index for the rows and the lower for the columns and again, with summation over repeated indices understood) for the spinors and the dirac matrices we find

$$\left(\bar{u}_{r_2}\right)_i \left(\gamma^{\nu'}\right)^i_j \left(\not{p}_2 - \not{k}_2 + m\right)^j_k \left(\gamma^{\mu'}\right)^k_l \left(v_{r_1}\right)^l \left(\bar{v}_{r_1}\right)_m$$

$$\times \left(\gamma^\mu\right)^m_n \left(\not{p}_2 - \not{k}_2 + m\right)^n_o \left(\gamma^\nu\right)^o_p \left(u_{r_2}\right)^p =$$

$$= \left(\gamma^{\nu'}\right)^i_j \left(\not{p}_2 - \not{k}_2 + m\right)^j_k \left(\gamma^{\mu'}\right)^k_l \left(v_{r_1}\right)^l \left(\bar{v}_{r_1}\right)_m$$

$$\times \left(\gamma^\mu\right)^m_n \left(\not{p}_2 - \not{k}_2 + m\right)^n_o \left(\gamma^\nu\right)^o_p \left(u_{r_2}\right)^p \left(\bar{u}_{r_2}\right)_i =$$

$$= \mathrm{Tr}\left\{ \gamma^{\nu'} \left(\not{p}_2 - \not{k}_2 + m\right) \gamma^{\mu'} v_{r_1} \bar{v}_{r_1} \gamma^\mu \left(\not{p}_2 - \not{k}_2 + m\right) \gamma^\nu u_{r_2} \bar{u}_{r_2} \right\}. \quad (5.60)$$

Summing over the polarizations of the fermions and of the photons we obtain

$$\sum_{r,s,r_1,r_2} |\mathcal{M}_1|^2 = e^4 Q^4 \left(-g_{\mu\mu'}\right)\left(-g_{\nu\nu'}\right) \frac{1}{[(p_2 - k_2)^2 - m^2]^2}$$

$$\times \mathrm{Tr}\left\{ \gamma^{\nu'} \left(\not{p}_2 - \not{k}_2 + m\right) \gamma^{\mu'} \left(\not{p}_1 - m\right) \right.$$

$$\left. \times \gamma^\mu \left(\not{p}_2 - \not{k}_2 + m\right) \gamma^\nu \left(\not{p}_2 + m\right) \right\} \quad (5.61)$$

The corresponding formulae needed for calculating cross sections and decay rates are given in Chap. 3.

Note that, when summing over the polarizations of the photon we have made the usual substitution

$$\sum_r \epsilon^{r*}_\mu(p) \epsilon^r_\nu(p) \rightarrow -g_{\mu\nu}. \quad (5.62)$$

This can safely be done in QED due to the fact that it is an abelian gauge theory. The photon always couples to a conserved current and the non-physical polarizations do not contribute to the physical observables. For QCD, which is non-abelian, the previous substitution must be done with care. We shall discuss this in the next subsection.

5.4.2 Non Abelian Gauge Theories

Consider the following QCD process

$$\bar{q}(p_1) q(p_2) \rightarrow g(p_3) g(p_4). \quad (5.63)$$

where $q\bar{q}$ stands for a quark-antiquark pair and g stands for gluon. For this process one has the same two diagrams from the previous section (with gluons substituting the photons and quarks substituting the electrons) plus a third one, that involves a triple gluon vertex (which is absent in QED). The complete set is shown in Fig. 5.2,

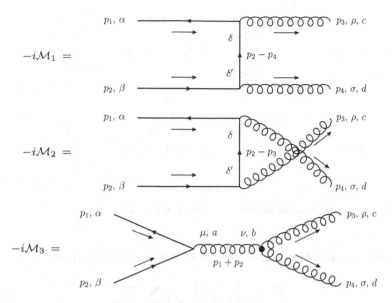

Fig. 5.2 Feynman diagrams for $q\bar{q}$ annihilation to a pair of gluons

where we have also included the corresponding colour labels and Lorentz indices. Due to the presence of the third diagram, if one makes the same substitution (5.62) for the gluon polarizations, one will obtain the wrong result. It is so, because one includes into the squared transition matrix non-physical polarization of the gluons (they are not coupled any more to conserved currents for this third diagram, thus the non-physical polarization do not vanish). There are two solutions. One can explicitly introduce the expressions for the gluon polarizations and only sum over the two physical ones (which can result rather tedious) or make the substitution (5.62) and introduce the Fadeev-Popov ghosts (at tree level). This is the approach that we shall present next.

Therefore, in order to get rid of the contributions of the non-physical polarizations of the gluons we will sum the ghost-antighost ($c\bar{c}$) contributions from Fig. 5.3 (diagram (A) **or** diagram (B), as they both lead to the same result).
Thus, the (correct result for the) total cross section will be given by

$$\boxed{\sigma(q\bar{q} \to gg) = \bar{\sigma}(q\bar{q} \to gg) + \sigma(q\bar{q} \to c\bar{c})}. \qquad (5.64)$$

The explicit expressions for the cross sections read

$$\bar{\sigma}(q\bar{q} \to gg) = \frac{1}{2\lambda^{1/2}(s, m_q^2, m_q^2)} \overline{\sum} \int dQ_2 \left| \mathcal{M}(q\bar{q} \to gg) \right|^2, \qquad (5.65)$$

$$\bar{\sigma}(q\bar{q} \to c\bar{c}) = \frac{1}{2\lambda^{1/2}(s, m_q^2, m_q^2)} \overline{\sum} \int dQ_2 \left| \mathcal{M}_A(q\bar{q} \to c\bar{c}) \right|^2, \qquad (5.66)$$

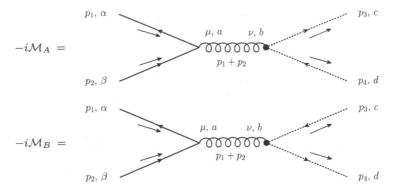

Fig. 5.3 Feynman diagrams for $q\bar{q}$ annihilation to ghost-antighost

where we have introduced the spin and colour ($N_c = 3$) averaged sum

$$\overline{\sum} \equiv \frac{1}{N_c^2} \frac{1}{(2j_q + 1)^2} \sum_{\text{pol.}} \sum_{\text{col.}}, \qquad (5.67)$$

and where

$$|\mathcal{M}(q\bar{q} \to gg)|^2 = |\mathcal{M}_1|^2 + |\mathcal{M}_2|^2 + |\mathcal{M}_3|^2 + 2\,\text{Re}\,\mathcal{M}_1^\dagger \mathcal{M}_2$$
$$+ 2\,\text{Re}\,\mathcal{M}_2^\dagger \mathcal{M}_3 + 2\,\text{Re}\,\mathcal{M}_1^\dagger \mathcal{M}_3. \qquad (5.68)$$

When calculating these cross sections one must remember that dQ_2 includes an identical particles $1/2$ symmetry factor for the cross section with gluons in the final state, whereas this factor does not exist for the cross section with ghosts in the final state (a ghost and an anti-ghost are not identical particles).

It is left as an exercise for the reader to explicitly check the following results (the needed Feynman rules in the Feynman gauge $\xi = 1$ are given in Fig. 5.4; the fermion-fermion-gluon vertex is proportional to the matrix elements $(\lambda^a)_{\alpha\beta}$, where λ^a are the Gell-Mann matrices, thus for the external fields the same Feynman rules from QED are valid except, fermions must also have a colour label α, β, etc. and gluon polarization vectors must have a, b, etc. labels).[4] For initial state massless quarks we have (summation over repeated colour indices must not be understood except if the symbol $\sum$ is explicitly introduced)

---

[4]For a complete set of Feynman rules for QCD (and the SM in general) see J. C. Romao and J. P. Silva, *A resource for signs and Feynman diagrams of the Standard Model*, Int. J. Mod. Phys. A **27** (2012) 1230025, http://arxiv.org/pdf/1209.6213.pdf

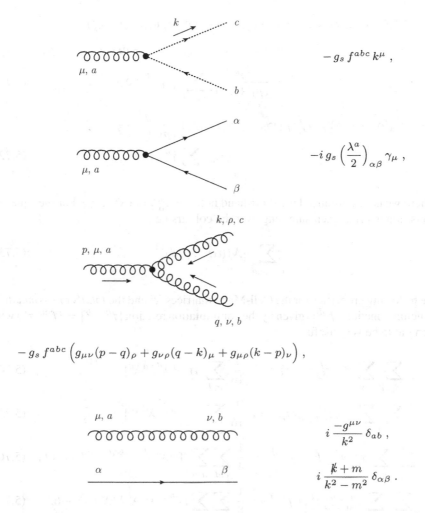

$$-g_s f^{abc} k^\mu \,,$$

$$-i\, g_s \left(\frac{\lambda^a}{2}\right)_{\alpha\beta} \gamma_\mu \,,$$

$$-g_s f^{abc} \left(g_{\mu\nu}(p-q)_\rho + g_{\nu\rho}(q-k)_\mu + g_{\mu\rho}(k-p)_\nu\right) \,,$$

$$i\,\frac{-g^{\mu\nu}}{k^2}\,\delta_{ab}\,,$$

$$i\,\frac{\slashed{k}+m}{k^2-m^2}\,\delta_{\alpha\beta}\,.$$

Fig. 5.4 Feynman rules needed for the $q\bar{q} \to gg$ process

$$-i\,\mathcal{M}_1(\alpha,\beta,c,d) = -i\,\epsilon^{\rho*}_{c,r_3}(p_3)\,\epsilon^{\sigma*}_{d,r_4}(p_4)\,\bar{v}^\alpha_{r_1}(p_1)\gamma_\rho$$
$$\times\,\frac{\slashed{p_2}-\slashed{p_4}}{(p_2-p_4)^2}\,\gamma_\sigma\,u^\beta_{r_2}(p_2)\,g_s^2 \sum_{\delta,\delta'} \tau^c_{\alpha\delta}\,\tau^d_{\delta'\beta}\,\delta_{\delta\delta'}, \qquad (5.69)$$

$$-i\,\mathcal{M}_2(\alpha,\beta,c,d) = -i\,\epsilon^{\rho*}_{c,r_3}(p_3)\,\epsilon^{\sigma*}_{d,r_4}(p_4)\,\bar{v}^\alpha_{r_1}(p_1)\gamma_\rho$$
$$\times\,\frac{\slashed{p_2}-\slashed{p_3}}{(p_2-p_3)^2}\,\gamma_\sigma\,u^\beta_{r_2}(p_2)\,g_s^2 \sum_{\delta,\delta'} \tau^d_{\alpha\delta}\,\tau^c_{\delta'\beta}\,\delta_{\delta\delta'}, \qquad (5.70)$$

$$-i \, \mathcal{M}_3(\alpha, \beta, c, d) = \epsilon^{\rho*}_{c,r_3}(p_3) \, \epsilon^{\sigma*}_{d,r_4}(p_4) \, \bar{v}^{\alpha}_{r_1}(p_1) \gamma_\mu u^{\beta}_{r_2}(p_2)(-g_s^2)$$

$$\times \left(g_{\nu\rho}(k + p_3)_\sigma + g_{\rho\sigma}(-p_3 + p_4)_\nu + g_{\sigma\nu}(-p_4 - k)_\rho \right)$$

$$\times \frac{-g^{\mu\nu}}{(p_1 + p_2)^2} \sum_{a,b} \tau^a_{\alpha\beta} \, f^{bcd} \, \delta_{ab} \,, \tag{5.71}$$

$$-i \, \mathcal{M}_A(\alpha, \beta, c, d) = \bar{v}^{\alpha}_{r_1}(p_1) \gamma_\mu \, u^{\beta}_{r_2}(p_2) \, (-g_s^2) \, \frac{-g^{\mu\nu}}{(p_1 + p_2)^2}$$

$$\times p_{3,\nu} \sum_{a,b} \tau^a_{\alpha\beta} \, f^{bcd} \, \delta_{ab} \,, \tag{5.72}$$

where we have introduced the short-hand notation $\tau^a_{\alpha\beta} = (\lambda^a/2)_{\alpha\beta}$. For the squared transition matrix, when summing over the colours i.e.,

$$\sum_{\alpha,\beta,c,d} \left| \mathcal{M}(\alpha, \beta, c, d) \right|^2 \,, \tag{5.73}$$

the following properties for the Gell-Mann matrices λ^a and the *totally antisymmetric* structure functions f^{abc} (given by the commutation relation $[\tau^a, \tau^b] = i f^{abc} \tau^c$) will turn out to be very useful

$$\sum_{\alpha,\beta,c,d} \sum_{\delta,\delta'} \tau^c_{\alpha\delta} \tau^d_{\delta\beta} \tau^d_{\beta\delta'} \tau^c_{\delta'\alpha} = \frac{1}{16} \sum_{c,d} \text{Tr}\{\lambda^c \lambda^d \lambda^d \lambda^c\} = \frac{16}{3} \,, \tag{5.74}$$

$$\sum_{\alpha,\beta,c,d} \sum_{\delta,\delta'} \tau^d_{\alpha\delta} \tau^c_{\delta\beta} \tau^d_{\beta\delta'} \tau^c_{\delta'\alpha} = \frac{1}{16} \sum_{c,d} \text{Tr}\{\lambda^d \lambda^c \lambda^d \lambda^c\} = -\frac{2}{3} \,, \tag{5.75}$$

$$\sum_{\alpha,\beta,c,d} \sum_{a,b} \tau^a_{\beta\alpha} \tau^b_{\alpha\beta} f^{bcd} f^{acd} = \frac{1}{4} \sum_{c,d} \sum_{a,b} \text{Tr}\{\lambda^a \lambda^b\} f^{bcd} f^{acd} = 12 \,, \tag{5.76}$$

$$\sum_{\alpha,\beta,c,d} \sum_{\delta,b} \tau^d_{\delta\alpha} \tau^b_{\alpha\beta} \tau^c_{\beta\delta} i \, f^{bcd} = \frac{1}{8} \sum_{c,d} \sum_b i f^{bcd} \, \text{Tr}\{\lambda^b \lambda^d \lambda^c\} = -6 \,. \tag{5.77}$$

We shall only perform the explicit calculation for the first squared amplitude:

$$-i \, \mathcal{M}_1(\alpha, \beta, c, d) = -i \, \epsilon^{\rho*}_{c,r_3}(p_3) \, \epsilon^{\sigma*}_{d,r_4}(p_4) \, \bar{v}^{\alpha}_{r_1}(p_1) \gamma_\rho$$

$$\times \frac{\not{p}_2 - \not{p}_4}{(p_2 - p_4)^2} \, \gamma_\sigma \, u^{\beta}_{r_2}(p_2) \, g_s^2 \sum_\delta \tau^c_{\alpha\delta} \tau^d_{\delta\beta}, \tag{5.78}$$

$$i \, \mathcal{M}^{\dagger}_1(\alpha, \beta, c, d) = i \, \epsilon^{\rho'}_{c,r_3}(p_3) \, \epsilon^{\sigma'}_{d,r_4}(p_4) \, \bar{u}^{\beta}_{r_2}(p_2) \gamma_{\sigma'}$$

$$\times \frac{\not{p}_2 - \not{p}_4}{(p_2 - p_4)^2} \, \gamma_{\rho'} \, v^{\alpha}_{r_1}(p_1) \, g_s^2 \sum_{\delta'} \tau^c_{\delta'\alpha} \tau^d_{\beta\delta'}. \tag{5.79}$$

Summing over the gluon and fermion polarizations

$$\sum_{r_i} |\mathcal{M}_1(\alpha, \beta, c, d)|^2 = (-g^{\rho\rho'})(-g^{\sigma\sigma'}) \frac{g_s^4}{t^2} \sum_{\delta,\delta'} \tau^c_{\alpha\delta} \tau^d_{\delta\beta} \tau^d_{\beta\delta'} \tau^c_{\delta'\alpha}$$

$$\times \operatorname{Tr}\{\gamma_{\sigma'}(\not{p}_2 - \not{p}_4)\gamma_{\rho'} \not{p}_1 \gamma_\rho (\not{p}_2 - \not{p}_4)\gamma_\sigma \not{p}_2\}. \quad (5.80)$$

Finally, summing over the colours, using (5.74–5.77), introducing the spin and colour average factors from (5.67), one obtains the following results

$$\overline{\sum}|\mathcal{M}_1|^2 = g_s^4 \frac{32\,u}{27\,t}, \quad (5.81)$$

$$\overline{\sum}|\mathcal{M}_2|^2 = g_s^4 \frac{32\,t}{27\,u}, \quad (5.82)$$

$$\overline{\sum}|\mathcal{M}_3|^2 = -4\,g_s^4 \frac{4\,t^2 + 3\,u\,t + 4\,u^2}{3\,s^2}, \quad (5.83)$$

$$\overline{\sum}|\mathcal{M}_A|^2 = -g_s^4 \frac{2\,t\,u}{3\,s^2}, \quad (5.84)$$

and for the crossed terms

$$\overline{\sum} 2\operatorname{Re}\mathcal{M}_1^\dagger\mathcal{M}_2 = g_s^4 \left(\frac{8\,s}{27}\right) \frac{s+t+u}{t\,u} = 0, \quad (5.85)$$

$$\overline{\sum} 2\operatorname{Re}\mathcal{M}_2^\dagger\mathcal{M}_3 = g_s^4 \frac{4\,(s+t-u)}{3\,s}, \quad (5.86)$$

$$\overline{\sum} 2\operatorname{Re}\mathcal{M}_1^\dagger\mathcal{M}_3 = g_s^4 \frac{4\,(s-t+u)}{3\,s}. \quad (5.87)$$

The final result for the total differential cross section is given by[5]

$$\boxed{\frac{d\sigma(q\bar{q} \to gg)}{dt} = \frac{\alpha_s^2\,\pi}{2\,s^2} \left[\frac{32}{27}\left(\frac{t^2+u^2}{t\,u}\right) - \frac{8}{3}\left(\frac{t^2+u^2}{s^2}\right) \right]}, \quad (5.88)$$

where we have introduced $\alpha_s = g_s^2/4\pi$, the strong coupling constant.

Further Reading

A. Pich, *Quantum Chromodynamics*, http://arxiv.org/pdf/hep-ph/9505231.pdf
G. Dissertori, I.G. Knowles, M. Schmelling, *Quantum Chromodynamics*
A. Pich, *The Standard Model of Electroweak Interactions*, http://arxiv.org/pdf/1201.0537v1.pdf

---

[5]The result from Peskin and Schroeder for this cross section seems not to be correct, however our result agrees (except a 1/2 factor which they did not include) with R.K. Ellis, W.J. Stirling, B.R. Webber, *QCD and Collider Physics*.

T. Muta, *Foundations of Quantum Chromodynamics*

J.C. Romao, J.P. Silva, A resource for signs and Feynman diagrams of the standard model, Int. J. Mod. Phys. A **27** (2012) 1230025, http://arxiv.org/pdf/1209.6213.pdf

R.K. Ellis, W.J. Stirling, B.R. Webber, *QCD and Collider Physics*

T.P. Cheng, L.F. Li, *Gauge Theory of Elementary Particle Physics*, (Oxford, 1984)

L.H. Ryder, *Quantum Field Theory*, (Cambridge University Press, 1985)

F. Mandl, G.P. Shaw, *Quantum Field Theory*

M. Kaku, *Quantum Field Theory: A Modern Introduction.*

M.E. Peskin, D.V. Schroeder, *An Introduction To Quantum Field Theory* (Addison-Wesley Publishing Company, San Francisco, 1995)

Chapter 6
Dimensional Regularization. Ultraviolet and Infrared Divergences

Abstract The cornerstone of Quantum Field Theory is renormalization. We shall speak more about in the next chapters. Before, it is necessary to discuss the method. The best and most simple is, of course, dimensional regularization (doesn't break the symmetries, doesn't violate the Ward Identities, preserves Lorentz invariance, etc.). When explained consistently, it becomes very simple and clear. Here, we shortly discuss ultraviolet (UV) and infrared (IR) divergences with a few examples. However, in Chap. 8, we shall extensively treat one-loop two and three-point functions and analyse many more examples of IR and UV divergences.

6.1 Master Integral

When calculating higher order quantum corrections (loop diagrams) one will find that, normally, the loop integrals are divergent. In order to make sense out of this divergent integrals and be able to properly define finite observables one has to regularize the divergent integrals (*make* them finite) and renormalize. In this chapter we shall treat the regularization procedure. In the next chapter we shall talk about renormalization.

In order to regularize the loop integrals, one can use different approaches i.e., cut-off, Wilson regularization, dimensional regularization, etc. This last approach is the one we shall treat in this chapter. It is the most adequate due to the fact that it preserves all the symmetries of the theory. It consists in considering that the space-time dimension is not 4 but $D = 4 + 2\epsilon$ and work in the limit $\epsilon \to 0$. With this consideration all loop-integrals are finite. The general D-dimensional integral that we will always relate to (as we shall shortly see in the next sections) is

$$J(D, \alpha, \beta, a^2) \equiv \int \frac{d^D k}{(2\pi)^D} \frac{(k^2)^\alpha}{(k^2 - a^2)^\beta}, \qquad (6.1)$$

where D is the number of space-time dimensions, as we already mentioned. We can easily demonstrate (see Appendix A) that it can be written in terms of the Euler Gamma function as

© Springer International Publishing Switzerland 2016
V. Ilisie, *Concepts in Quantum Field Theory*,
UNITEXT for Physics, DOI 10.1007/978-3-319-22966-9_6

$$J(D, \alpha, \beta, a^2) = \frac{i}{(4\pi)^{D/2}} (a^2)^{D/2} (-a^2)^{\alpha-\beta} \frac{\Gamma(\beta - \alpha - D/2)\Gamma(\alpha + D/2)}{\Gamma(\beta)\Gamma(D/2)}.$$

(6.2)

For z a complex number with $Re(z) > 0$ we have the following property of $\Gamma(z)$

$$\Gamma(z + 1) = z\Gamma(z).$$

(6.3)

For $n = 0, 1, 2, 3, \ldots$, it takes the form of the usual factorial function

$$\Gamma(n + 1) = n!.$$

(6.4)

The function $\Gamma(z)$ has simple poles at $z = 0, -1, -2, \ldots$ In the region $-1 < \epsilon < 0$ with $\epsilon \in \mathbb{R}$ and $|\epsilon| \ll 1$, we have the following Laurent expansion up to $O(\epsilon)$

$$\Gamma(-\epsilon) = -\frac{1}{\epsilon} - \gamma_E + O(\epsilon),$$

(6.5)

where $\gamma_E = 0.57721\ldots$ is the Euler-Mascheroni constant. Taking $D = 4 + 2\epsilon$ with $-1 < \epsilon < 0$, $|\epsilon| \ll 1$ (UV-divergent integrals in $D = 4$ dimensions are convergent in $D < 4$ dimensions) and power-expanding in ϵ, it is straightforward to find the expression for the UV divergent function $J(D, 0, 2, a^2)$:

$$J(D, 0, 2, a^2) = \frac{-i}{(4\pi)^2} \left(\frac{1}{\epsilon} + \gamma_E - \ln(4\pi) + \ln(a^2) + O(\epsilon) \right).$$

(6.6)

It will be useful to define the following quantity:

$$\frac{1}{\hat{\epsilon}} \equiv \frac{1}{\epsilon} + \gamma_E - \ln(4\pi).$$

(6.7)

Because a in (6.6) has energy dimensions within the logarithm, we multiply and divide the RHS of (6.6) by $\mu^{2\epsilon}$, with μ a parameter with energy dimensions (called the renormalization scale), and use the expansion

$$\mu^{-2\epsilon} = 1 - 2\epsilon \ln(\mu) + O(\epsilon^2),$$

(6.8)

to finally obtain the following simple expression for $J(D, 0, 2, a^2)$

$$J(D, 0, 2, a^2) = \frac{-i}{(4\pi)^2} \mu^{2\epsilon} \left[\frac{1}{\hat{\epsilon}} + \ln\left(\frac{a^2}{\mu^2}\right) \right] + O(\epsilon).$$

(6.9)

We can easily relate all the UV divergent integrals to this one by using the recursion properties of the Gamma function so practically **one does not have to integrate ever again over the four-momentum** (except for some special cases where IR divergences are also present). Some useful results are presented next.

6.2 Useful Results

Using the Gamma function recursion properties we can find very useful the following relations

$$J(D, 0, 1, a^2) = \frac{a^2}{D/2 - 1} J(D, 0, 2, a^2)$$

$$J(D, 1, 1, a^2) = \frac{a^4}{D/2 - 1} J(D, 0, 2, a^2)$$

$$J(D, 1, 2, a^2) = \frac{a^2 D}{D - 2} J(D, 0, 2, a^2)$$

$$J(D, 2, 2, a^2) = \frac{a^4(D + 2)}{D - 2} J(D, 0, 2, a^2)$$

$$J(D, 1, 3, a^2) = \frac{D}{4} J(D, 0, 2, a^2)$$

(6.10)

As we shall see, all these results will be very useful in Chap. 8. One can construct similar recursion relations for any values of α and β of our function $J(D, \alpha, \beta, a^2)$ with no additional complications.

It is worth mentioning that the previous relations are general, meaning that they are valid for any dimension D ($D < 4$ or $D > 4$). This can turn out to be useful also when treating IR divergences (IR divergent integrals in $D = 4$ dimensions are convergent in $D > 4$ dimensions).

An example of integral that is finite in $D = 4$ dimensions is the following:

$$J(4, 0, 3, a^2) = \frac{-i}{32\pi^2} \frac{1}{a^2}.$$

(6.11)

When IR divergences are present we have to be a little bit more careful (with this integral) as we shall shortly see with an explicit example. Some other useful generic results in D dimensions for the Dirac gamma matrices are the following:

$$g^{\mu\nu} g_{\mu\nu} = D I_D,$$
$$\gamma^\mu \gamma^\nu \gamma_\mu = (2 - D)\gamma^\nu,$$
$$\gamma^\mu \gamma^\nu \gamma^\sigma \gamma_\mu = 4g^{\nu\sigma} I_D + (D - 4)\gamma^\nu \gamma^\sigma,$$
$$\gamma^\mu \gamma^\nu \gamma^\sigma \gamma^\rho \gamma_\mu = -2\gamma^\rho \gamma^\sigma \gamma^\nu - (D - 4)\gamma^\nu \gamma^\sigma \gamma^\rho,$$

(6.12)

where I_D is the identity matrix in D dimensions. The standard convention is to take its trace equal to four:

$$Tr\{I_D\} = 4. \tag{6.13}$$

Many more relations of this type involving gamma matrices are present all over the literature. Some other interesting results that involve the function $J(D, \alpha, \beta, a^2)$ are the following:

$$\int \frac{d^D k}{(2\pi)^D} \frac{(k^2)^\alpha k^\mu}{(k^2 - a^2)^\beta} = 0, \tag{6.14}$$

$$\int \frac{d^D k}{(2\pi)^D} \frac{(k^2)^\alpha k^\mu k^\nu}{(k^2 - a^2)^\beta} = \frac{g^{\mu\nu}}{D} J(D, \alpha + 1, \beta, a^2) \tag{6.15}$$

Using the last result we can straightforwardly deduce another relation:

$$\int \frac{d^D k}{(2\pi)^D} \frac{(k \cdot p)^2}{(k^2 - a^2)^\beta} = \int \frac{d^D k}{(2\pi)^D} \frac{k^\mu k^\nu p_\mu p_\nu}{(k^2 - a^2)^\beta}$$

$$= \int \frac{d^D k}{(2\pi)^D} \frac{p_\mu p_\nu}{(k^2 - a^2)^\beta} \frac{g^{\mu\nu} k^2}{D}$$

$$= \frac{1}{D} \int \frac{d^D k}{(2\pi)^D} \frac{k^2 p^2}{(k^2 - a^2)^\beta}$$

$$= \frac{p^2}{D} J(D, 1, \beta, a^2). \tag{6.16}$$

Thus, we have found the following equality in D dimensions

$$\int \frac{d^D k}{(2\pi)^D} \frac{(k \cdot p)^2}{(k^2 - a^2)^\beta} = \int \frac{d^D k}{(2\pi)^D} \frac{(p^2/D)k^2}{(k^2 - a^2)^\beta}. \tag{6.17}$$

The reader is highly encouraged to find the generalization of this result for $(k \cdot p)^\alpha$ with α an arbitrary positive integer.

6.3 Example: Cancellation of UV Divergences

The following is a simple but very useful example of how the divergent parts of an integral cancel and lead to a finite final result. Consider

$$G^{\mu\nu} = \int \frac{d^D k}{(2\pi)^D} \frac{4k^\mu k^\nu - g^{\mu\nu} k^2}{[k^2 - a^2]^3}. \tag{6.18}$$

The calculation is straightforward. Using (6.10) we obtain

$$
\begin{aligned}
G^{\mu\nu} &= g^{\mu\nu}\left(\frac{4}{D} - 1\right) J(D, 1, 3, a^2) \\
&= g^{\mu\nu}\left(\frac{4}{D} - 1\right)\frac{D}{4} J(D, 0, 2, a^2) \\
&= g^{\mu\nu}\left(-\frac{\epsilon}{2}\right) J(D, 0, 2, a^2) + O(\epsilon^2) \\
&= g^{\mu\nu}\frac{i}{32\pi^2} + O(\epsilon).
\end{aligned}
\tag{6.19}
$$

Taking $D \to 4$ we obtain:

$$
\boxed{G^{\mu\nu} \equiv \int \frac{d^D k}{(2\pi)^D} \frac{4k^\mu k^\nu - g^{\mu\nu}k^2}{[k^2 - a^2]^3} = g^{\mu\nu}\frac{i}{32\pi^2}.}
\tag{6.20}
$$

It is a common mistake to think that because the final result is finite in 4 dimensions we could have made directly the substitution $k^\mu k^\nu \to g^{\mu\nu}k^2/4$ instead of $k^\mu k^\nu \to g^{\mu\nu}k^2/D$. Wrong! This integral consists in the sum of two parts that diverge in 4 dimensions. Only after these two parts are summed, the final result turns out to be finite. Thus, the substitution $k^\mu k^\nu \to g^{\mu\nu}k^2/D$ is only valid if it gives rise to a finite result. In our case both parts are finite for $D < 4$ (and not $D = 4$) dimensions and therefore, we must maintain $D = 4 + 2\epsilon$ until the sum of both parts is performed.

6.4 Feynman Parametrization

Usually we don't find simple propagators as in (6.1), therefore we have to perform some manipulations over the denominators. The standard procedure is using the Feynman parametrization:

$$
\frac{1}{A^\alpha B^\beta} = \frac{\Gamma(\alpha + \beta)}{\Gamma(\alpha)\Gamma(\beta)} \int_0^1 dx \frac{x^{\alpha-1}(1 - x)^{\beta-1}}{[Ax + B(1 - x)]^{\alpha+\beta}},
\tag{6.21}
$$

$$
\frac{1}{A^\alpha B^\beta C^\gamma} = \frac{\Gamma(\alpha + \beta + \gamma)}{\Gamma(\alpha)\Gamma(\beta)\Gamma(\gamma)} \int_0^1 dx \int_0^1 dy\, x
$$
$$
\times \frac{(xy)^{\alpha-1}[x(1 - y)]^{\beta-1}(1 - x)^{\gamma-1}}{[Axy + Bx(1 - y) + C(1 - x)]^{\alpha+\beta+\gamma}}.
\tag{6.22}
$$

For $\alpha = \beta = \gamma = 1$ we simply get:

$$\frac{1}{AB} = \int_0^1 dx \frac{1}{[Ax + B(1-x)]^2}, \tag{6.23}$$

$$\frac{1}{ABC} = \int_0^1 dx \int_0^1 dy \frac{2x}{[Axy + Bx(1-y) + C(1-x)]^3}. \tag{6.24}$$

An useful alternative for (6.24) is the following:

$$\frac{1}{ABC} = 2 \int_0^1 dx \int_0^1 dy \int_0^1 dz \frac{\delta(1-x-y-z)}{[Ax + By + Cz]^3}$$

$$= \int_0^1 dy \int_0^{1-y} dz \frac{2}{[A(1-y-z) + By + Cz]^3}. \tag{6.25}$$

For a generic n-point function one can use the generalized Feynman parametrization given by:

$$\frac{1}{A_1^{\alpha_1} A_2^{\alpha_2} \dots A_n^{\alpha_n}} = \frac{\Gamma(\alpha)}{\Gamma(\alpha_1) \Gamma(\alpha_2) \dots \Gamma(\alpha_n)} \int_0^1 dx_1 \int_0^1 dx_2 \dots \int_0^1 dx_n$$

$$\times \frac{x_1^{\alpha_1-1} x_2^{\alpha_2-1} \dots x_n^{\alpha_n-1}}{\left(x_1 A_1 + x_2 A_2 + \dots + x_n A_n\right)^\alpha}$$

$$\times \delta(1 - x_1 - x_2 - \dots - x_n), \tag{6.26}$$

where we have defined $\alpha \equiv \alpha_1 + \alpha_2 + \dots + \alpha_n$. Integrating over the δ-function, we find

$$\frac{1}{A_1^{\alpha_1} A_2^{\alpha_2} \dots A_n^{\alpha_n}} = \frac{\Gamma(\alpha)}{\Gamma(\alpha_1) \Gamma(\alpha_2) \dots \Gamma(\alpha_n)}$$

$$\times \int_0^1 dx_1 \int_0^{1-x_1} dx_2 \dots \int_0^{1-x_1-x_2-\dots-x_{n-2}} dx_{n-1}$$

$$\times \frac{x_1^{\alpha_1-1} x_2^{\alpha_2-1} \dots x_{n-1}^{\alpha_{n-1}-1} (1 - x_1 - \dots - x_{n-1})^{\alpha_n-1}}{\left(x_1 A_1 + x_2 A_2 + \dots + x_{n-1} A_{n-1} + (1 - x_1 - \dots - x_{n-1}) A_n\right)^\alpha}$$

$$\tag{6.27}$$

Next we shall see a few rather simple examples of how these parametrizations can be used. However in Chap. 8, we shall use these parametrizations to calculate more complicated UV and IR divergent integrals.

It is worth mentioning that one can consider the analytical prolongation to the complex plane and use this last parametrization (6.27) for non-integer powers of propagators. They appear when one wants integrate over the four-momentum, logarithmic functions that depend on the four-momentum. At the end of Chap. 8 we shall also take a simple two-loop example to see how this can be done.

6.5 Example: UV Pole

Let's consider the following integral:

$$I^\mu \equiv \int \frac{d^D k}{(2\pi)^D} \frac{k^\mu}{((p+k)^2 - m^2)k^2}.$$ (6.28)

Using the parametrization in (6.23) with $A = (p+k)^2 - m^2$ and $B = k^2$ we obtain

$$\frac{1}{((p+k)^2 - m^2)k^2} = \int_0^1 dx \frac{1}{[((p+k)^2 - m^2)x + k^2(1-x)]^2}$$

$$= \int_0^1 dx \frac{1}{[(k+px)^2 - a^2]^2},$$ (6.29)

where we have defined $a^2 \equiv -p^2 x(1-x) + m^2 x$. Therefore I^μ takes the form

$$\begin{aligned}
I^\mu &= \int_0^1 dx \int \frac{d^D k}{(2\pi)^D} \frac{k^\mu}{[(k+px)^2 - a^2]^2} \\
&= \int_0^1 dx \int \frac{d^D k}{(2\pi)^D} \frac{k^\mu - xp^\mu}{[k^2 - a^2]^2} \\
&= -p^\mu \int_0^1 x \, dx \, J(D, 0, 2, a^2) \\
&= p^\mu \frac{i}{(4\pi)^2} \mu^{2\epsilon} \int_0^1 x \, dx \left[\frac{1}{\hat{\epsilon}} + \ln\left(\frac{a^2}{\mu^2}\right) \right] \\
&= p^\mu \frac{i}{(4\pi)^2} \mu^{2\epsilon} \left[\frac{1}{2\hat{\epsilon}} + \int_0^1 x \, dx \ln\left(\frac{-p^2 x(1-x) + m^2 x}{\mu^2}\right) \right]
\end{aligned}$$ (6.30)

(to get to the second line we shifted the integration variable $k \to k - x\,p$).

6.6 Example: IR Poles

Consider the following integral

$$I = \int \frac{d^D k}{(2\pi)^D} \frac{1}{k^2(k+p_2)^2(k-p_3)^2},$$ (6.31)

with $p_2^2 = p_3^2 = 0$. Thus our denominator in the limit of small k behaves like

$$\frac{1}{k^2(k^2 + 2p_2 \cdot k)(k^2 - 2p_3 \cdot k)} \xrightarrow{k^2 \ll} \frac{1}{k^2(2p_2 \cdot k)(2p_3 \cdot k)}. \tag{6.32}$$

We can observe two types of infrared divergences present in the denominator of (6.32):

      1. $k \to 0$,   which is called soft divergence,

      2. $p_3 \cdot k$ or $p_2 \cdot k \to 0$,   which is called collinear divergence.   (6.33)

The integral that we are treating here will turn out to be only IR divergent, without UV divergences (a case where we find both types of divergences will be treated in Chap. 8). In order to treat these IR divergences, we shall consider the space-time dimensions to be $D = 4 + 2\epsilon'$ with $\epsilon' \in \mathbb{R}$, $\epsilon' > 0$ and $\epsilon' \ll 1$ (the IR divergent integrals in 4 dimensions are convergent in $D > 4$ dimensions, as we have already mentioned). Using the Feynman parametrization (6.24) with $A = k^2 + 2p_2 \cdot k$, $B = k^2 - 2p_3 \cdot k$ and $C = k^2$, we find

$$\begin{aligned} I &= \int_0^1 dx \int_0^1 dy \int \frac{d^D k}{(2\pi)^D} \frac{2x}{[(k + p_2 xy - p_3 x(1-y))^2 - a^2]^3} \\ &= \int_0^1 dx \int_0^1 dy \int \frac{d^D k}{(2\pi)^D} \frac{2x}{[k^2 - a^2]^3} \\ &= 2 \int_0^1 dx \int_0^1 dy\, x\, J(D, 0, 3, a^2), \end{aligned} \tag{6.34}$$

where $a^2 = -2(p_2 \cdot p_3) x^2 y(1-y)$. Taking $D \to 4$ as in (6.11) would give rise to an infinity when integrating over the Feynman parameters, as we will see in a moment. Let's write $J(D, 0, 3, a^2)$ in the form of (6.2):

$$\begin{aligned} I &= \frac{-i}{(4\pi)^{D/2}} \Gamma(3 - D/2) \int_0^1 dx \int_0^1 dy\, x (a^2)^{D/2-3} \\ &= \frac{-i}{(4\pi)^{D/2}} \Gamma(3 - D/2)(-2p_2 \cdot p_3)^{D/2-3} \int_0^1 dx\, x^{D-5} \\ &\qquad\qquad\qquad \times \int_0^1 dy\, y^{D/2-3}(1-y)^{D/2-3}. \end{aligned} \tag{6.35}$$

Using the Euler Beta function (A.17) we obtain the following result for our integral:

$$I = \frac{-i}{(4\pi)^{D/2}}\Gamma(3 - D/2)(-2p_2 \cdot p_3)^{D/2-3}\frac{\Gamma(D-4)}{\Gamma(D-3)}\frac{\Gamma(D/2-2)\Gamma(D/2-2)}{\Gamma(D-4)}$$

$$= \frac{-i}{(4\pi)^{D/2}}\frac{\Gamma(3-D/2)}{\Gamma(D-3)}(-2p_2 \cdot p_3)^{D/2-3}\Gamma(D/2-2)\Gamma(D/2-2)$$

$$= \frac{-i}{(4\pi)^2}(-2p_2 \cdot p_3)^{-1}(4\pi)^{-\epsilon'}(-2p_2 \cdot p_3)^{\epsilon'}\frac{\Gamma(1-\epsilon')}{\Gamma(1+2\epsilon')}\Gamma(\epsilon')\Gamma(\epsilon')$$

$$= \frac{i}{(4\pi)^2}\frac{1}{2p_2 \cdot p_3}\left(\frac{-2p_2 \cdot p_3}{4\pi}\right)^{\epsilon'}\frac{\Gamma(1-\epsilon')}{\Gamma(1+2\epsilon')}\Gamma^2(\epsilon'). \tag{6.36}$$

Of course, one could further expand using:

$$\Gamma(\epsilon') = \frac{1}{\epsilon'} - \gamma_E + \frac{1}{12}(\pi^2 + 6\gamma_E^2)\epsilon' + O(\epsilon'^2), \tag{6.37}$$

$$\Gamma(1 \mp \epsilon') = 1 \pm \gamma_E\epsilon' + \frac{1}{12}(\pi^2 + 6\gamma_E^2)\epsilon'^2 + O(\epsilon'^3), \tag{6.38}$$

$$(-1)^{\epsilon'} = e^{\pm i\pi\epsilon'} = 1 \pm i\pi\epsilon' - \frac{\pi^2\epsilon'^2}{2}. \tag{6.39}$$

Thus, we observe that our result (6.36) is proportional to IR poles of the form $1/\epsilon'$ and $1/\epsilon'^2$. This is why taking the limit $D \to 4$ as in (6.11), in (6.34) would have been wrong.

One could argue that in the first example we also have a propagator of the type $\sim 1/k^2$, which goes to infinity as k goes to zero. Therefore one should find IR divergences in this case also. It turns out, however, that it is not the case. It is clear that when integrating over x, the expression (6.30) does not diverge. This expression is divergent if $m^2 = p^2 = 0$, however this specific case will be treated in Chap. 8. Thus, any potential IR divergence must be treated carefully. As we can see, after doing the calculation, the IR divergence might actually not be there.

Further Reading

A. Pich, *The Standard Model of Electroweak Interactions*. http://arxiv.org/pdf/1201.0537v1.pdf
A. Pich, *Class Notes on Quantum Field Theory*. http://eeemaster.uv.es/course/view.php?id=6
M. Kaku, *Quantum Field Theory: A Modern Introduction*
M. Srednicki, *Quantum Field Theory*
M.E. Peskin, D.V. Schroeder, *An Introduction to Quantum Field Theory* (Addison-Wesley Publishing Company, San Francisco, 1995)
K. Kannike, *Notes on Feynman Parametrisation and the Dirac Delta Function*. http://www.physic. ut.ee/~kkannike/english/science/physics/notes/feynman_param.pdf
S. Pokorsky, *Gauge Field Theories*
L.H. Ryder, *Quantum Field Theory* (Cambridge University Press, Cambridge, 1985)
T.P. Cheng, L.F. Li, *Gauge Theory of Elementary Particle Physics* (Oxford University Press, Oxford, 1984)
F. Mandl, G.P. Shaw, *Quantum Field Theory*

Chapter 7
QED Renormalization

Abstract This chapter is intended to present the standard renormalization algorithm to all orders in perturbation theory in a simple transparent manner. The calculations are made as explicit as possible, in order to clarify many of the confusions that may arise when treating this topic. We will restrict our attention to the most academic example, which is (in my opinion) the QED Lagrangian. We shall use the dimensional regularization approach in $D = 4 + 2\epsilon$ dimensions ($\epsilon < 0$) in order to regularise the UV divergences (as it is nicely explained in the previous chapter). Afterwards we shall take an explicit one-loop example and introduce the two usual renormalization schemes, the $\overline{MS}$ and the on-shell scheme. A few clarifying notions on tadpoles are also given at the end of the chapter.

7.1 QED Lagrangian

The QED Lagrangian written in terms of the *bare* (non physical) fields, masses, coupling constant and gauge parameter reads

$$
\mathcal{L}_{QED} = -\frac{1}{4} F_{\mu\nu}^{(0)} F_{(0)}^{\mu\nu} - \frac{1}{2\xi_{(0)}} \left(\partial_\mu A_{(0)}^\mu \right)^2 + i \overline{\psi}^{(0)} \gamma^\mu \partial_\mu \psi^{(0)}
$$
$$
- m_0 \overline{\psi}^{(0)} \psi^{(0)} - e_0 Q A_\mu^{(0)} \overline{\psi}^{(0)} \gamma^\mu \psi^{(0)} \qquad (7.1)
$$

When including loop corrections the bare parameters will be replaced systematically by physical parameters of the renormalized theory as we shall see in this section. We start by analysing the perturbative corrections to all orders to propagators, vertex, and fields.

7.2 Fermionic Propagator, Mass and Field Renormalization

The relevant diagrams for renormalization are the one particle irreducible diagrams. A one particle irreducible diagram (**1PI**) is one that, by cutting an internal line is no longer connected. The 1PI fermionic self-energy diagrams that are needed for the

© Springer International Publishing Switzerland 2016
V. Ilisie, *Concepts in Quantum Field Theory*,
UNITEXT for Physics, DOI 10.1007/978-3-319-22966-9_7

fermionic propagator correction are schematically shown below:

$$i\,\Sigma(\not{p}, \zeta_0) =$$

where we have introduced the short-hand notation $\zeta_0 \equiv (m_0,\ \alpha_0,\ \xi_0)$ for the bare parameters. For the renormalized physical parameters we shall use $\zeta \equiv (m,\ \alpha,\ \xi)$. The fully dressed propagator is given by the *Dyson summation*:

$$i\,S(\not{p}, \zeta_0) =$$

Explicitly the previous sum reads:

$$i\,S(\not{p}, \zeta_0) = i\,S^{(0)}(\not{p}, \zeta_0) + i\,S^{(0)}(\not{p}, \zeta_0)\left(i\,\Sigma(\not{p}, \zeta_0)\right)i\,S^{(0)}(\not{p}, \zeta_0) + \cdots,$$

(7.2)

where $i\,S^{(0)}(\not{p}, \zeta_0) = \dfrac{i}{\not{p} - m_0}$ is the fermionic tree-level propagator. We obtain

$$S(\not{p}, \zeta_0) = \frac{1}{\not{p} - m_0} - \frac{\Sigma(\not{p}, \zeta_0)}{(\not{p} - m_0)^2} + \cdots = \frac{1}{\not{p} - m_0 + \Sigma(\not{p}, \zeta_0)}. \qquad (7.3)$$

We will now define the Z_2 renormalization constant that relates the bare (non-renormalized, UV-divergent) propagator $S(\not{p}, \zeta_0)$ with the renormalized one $S_R(\not{p}, \zeta)$ that contains only the regular (renormalized, UV finite) parts of the self-energy (denoted as Σ^R) and that depends on the physical (renormalized) parameters m and α and ξ:

$$S(\not{p}, \zeta_0) = \frac{1}{\not{p} - m_0 + \Sigma(\not{p}, \zeta_0)} \equiv Z_2\,S_R(\not{p}, \zeta) = \frac{Z_2}{\not{p} - m + \Sigma^R(\not{p}, \zeta)}. \qquad (7.4)$$

For the mass renormalization the standard procedure is to define the following renormalization constants:

$$m \equiv m_0 + \delta m \equiv Z_2\,Z_4^{-1}\,m_0. \qquad (7.5)$$

Using (7.4) we are now able to relate the renormalized fermionic fields $\psi(x)$ with the bare ones $\psi^{(0)}(x)$ with the help of the time-ordered product:

$$T\left(\psi^{(0)}(x), \overline{\psi}^{(0)}(y)\right) \sim \frac{i}{\not{p} - m_0 + \cdots} = \frac{i\,Z_2}{\not{p} - m + \cdots} \sim Z_2\,T\left(\psi(x), \overline{\psi}(y)\right).$$

(7.6)

We obtain thus, the following relations:

$$\psi^{(0)}(x) = Z_2^{1/2}\psi(x), \qquad \overline{\psi}^{(0)}(x) = Z_2^{1/2}\overline{\psi}(x). \tag{7.7}$$

Next we move on to the analysis of the bosonic propagator and field renormalization.

7.3 Bosonic Propagator and Field Renormalization

The 1PI diagrams that contribute to the photon self energy are given by:

$$i\,\Pi^{\mu\nu}(q^2, \varsigma_0) =$$

Due to gauge invariance $q_\mu \Pi^{\mu\nu} = q_\nu \Pi^{\mu\nu} = 0$, therefore these 1PI diagrams must necessarily have the form

$$\Pi^{\mu\nu}(q^2) = (-g^{\mu\nu}q^2 + q^\mu q^\nu)\,\Pi(q^2), \tag{7.8}$$

with $\Pi(q^2)$ a Lorentz scalar. The Dyson summation for the photon propagator is then given by:

$$i\,D^{\mu\nu}(q^2, \varsigma_0) =$$

The explicit expression for the previous sum is:

$$i D^{\mu\nu}(q^2, \varsigma_0) = i D_{(0)}^{\mu\nu}(q^2, \varsigma_0) + i D_{(0)}^{\mu\lambda}(q^2, \varsigma_0)\Big(i \Pi_{\lambda\rho}(q^2, \varsigma_0)\Big) i D_{(0)}^{\rho\nu}(q^2, \varsigma_0) + \cdots,$$

$$\tag{7.9}$$

where $i D_{(0)}^{\mu\nu}(q^2, \varsigma_0) = i\,\dfrac{-g^{\mu\nu} + (1-\varsigma_0)q^\mu q^\nu/q^2}{q^2}$ is the tree-level photon propagator. The second term of this sum gives:

$$D_{(0)}^{\mu\lambda}(q^2, \varsigma_0)\,\Pi_{\lambda\rho}(q^2, \varsigma_0)\,D_{(0)}^{\rho\nu}(q^2, \varsigma_0) = \frac{1}{q^2}\Big(-g^{\mu\nu} + \frac{q^\mu q^\nu}{q^2}\Big)\Pi(q^2, \varsigma_0). \tag{7.10}$$

Therefore the complete Dyson summation is finally given by:

$$
\begin{aligned}
D^{\mu\nu}(q^2,\zeta_0) &= \frac{1}{q^2}\left\{\left(-g^{\mu\nu}+\frac{q^{\mu}q^{\nu}}{q^2}\right)\left(1-\Pi(q^2,\zeta_0)+\cdots\right)-\xi_0\frac{q^{\mu}q^{\nu}}{q^2}\right\} \\
&= \frac{1}{q^2}\left\{\left(-g^{\mu\nu}+\frac{q^{\mu}q^{\nu}}{q^2}\right)\frac{1}{1+\Pi(q^2,\zeta_0)}-\xi_0\frac{q^{\mu}q^{\nu}}{q^2}\right\}.
\end{aligned} \tag{7.11}
$$

Now, as we did before, we relate $D^{\mu\nu}(q^2,\zeta_0)$ with the renormalized propagator $D_R^{\mu\nu}(q^2,\zeta)$ by defining another renormalization constant Z_3 as

$$
D^{\mu\nu}(q^2,\zeta_0) \equiv Z_3 D_R^{\mu\nu}(q^2,\zeta) = \frac{Z_3}{q^2}\left\{\left(-g^{\mu\nu}+\frac{q^{\mu}q^{\nu}}{q^2}\right)\frac{1}{1+\Pi_R(q^2,\zeta)}-\xi\frac{q^{\mu}q^{\nu}}{q^2}\right\},
\tag{7.12}
$$

where Π_R is the regular part of the self-energy function from (7.8). Thus we get to the following relations:

$$
\xi_0 = Z_3\,\xi, \qquad \frac{1}{1+\Pi(q^2,\zeta_0)} = \frac{Z_3}{1+\Pi_R(q^2,\zeta)}. \tag{7.13}
$$

Again, using the time-ordered product we obtain the expression for the renormalized photon field

$$
A_0^{\mu}(x) = Z_3^{1/2}A^{\mu}(x). \tag{7.14}
$$

The only thing we have left to complete the renormalization procedure is to analyse the vertex correction and the coupling constant renormalization.

7.4 Vertex Correction

The vertex correction to all orders in perturbation theory is schematically given by:

$$
-i\,e_0\,Q\,\Lambda^{\mu}(p_i,\zeta_0) =
$$

The complete vertex comes from summing the tree level one and the previous corrections:

$$-i\,e_0\,Q\,\Gamma^\mu(p_i,\zeta_0) =$$

Simplifying the $-i\,e_0\,Q$ common factor, this sum simply reads:

$$\Gamma^\mu(p_i,\zeta_0) = \gamma^\mu + \Lambda^\mu(p_i,\zeta_0). \tag{7.15}$$

The Z_1 constant that renormalizes the vertex is defined as follows:

$$\Gamma^\mu(p_i,\zeta_0) \;=\; \frac{1}{Z_1}\,\Gamma_R^\mu(p_i,\zeta) \;=\; \frac{1}{Z_1}\Big(\gamma^\mu + \Lambda_R^\mu(p_i,\zeta)\Big). \tag{7.16}$$

where Γ_R and Λ_R are (using the same notation as previously) regular (UV-finite) quantities. With the previous definition we can relate the bare coupling e_0 with the renormalized one:

$$e_0\,\Gamma^\mu(p_i,\zeta_0) \;=\; \frac{e_0}{Z_1}\,\Gamma_R^\mu(p_i,\zeta) \;\equiv\; e'\,\Gamma_R^\mu(p_i,\zeta). \tag{7.17}$$

We could be tempted to think that e' is the renormalized coupling constant but that's not really true. We must carefully look at the vertex diagram and realize that we have three external fields that also need renormalization. We have a bosonic field, therefore we must multiply e' by $Z_3^{1/2}$ and we have two fermion fields, therefore we need to also multiply by a Z_2 term. The renormalized coupling is then given by:

$$e \equiv \frac{e_0}{Z_1} Z_2 Z_3^{1/2} = e'\,Z_2\,Z_3^{1/2}. \tag{7.18}$$

We can demonstrate that the gauge invariance of the Lagrangian implies $\mathbf{Z_1} = \mathbf{Z_2}$. Thus, we finally obtain the simple relations between e and α (where $\alpha = e^2/4\pi$) and the correspondent bare quantities

$$e = e_0\,Z_3^{1/2} \;\Rightarrow\; \alpha = Z_3\,\alpha_0. \tag{7.19}$$

7.5 Renormalization to All Orders

To sum up, we have found the following relations between the bare and renormalized quantities

$$\boxed{\begin{aligned} \psi^{(0)} &= Z_2^{1/2}\psi, & A_0^\mu &= Z_3^{1/2} A^\mu, & e_0 &= Z_3^{-1/2} e, \\ \xi_0 &= Z_3\xi, & m_0 &= m - \delta m = Z_4 Z_2^{-1} m. \end{aligned}} \tag{7.20}$$

Now we can re-write the original Lagrangian in terms of the renormalized physical quantities using the Z_i renormalization constants that we have defined in the previous sections:

$$
\begin{aligned}
\mathcal{L}_{QED} = {}& -\frac{1}{4} F^{(0)}_{\mu\nu} F^{\mu\nu}_{(0)} - \frac{1}{2\xi_{(0)}} \left(\partial_\mu A^\mu_{(0)} \right)^2 + i\overline{\psi}^{(0)} \gamma^\mu \partial_\mu \psi^{(0)} \\
& - m_0 \overline{\psi}^{(0)} \psi^{(0)} - e_0 Q A^{(0)}_\mu \overline{\psi}^{(0)} \gamma^\mu \psi^{(0)} \\
= {}& -Z_3 \frac{1}{4} F_{\mu\nu} F^{\mu\nu} - \frac{1}{2\xi} (\partial_\mu A^\mu)^2 + Z_2 i\overline{\psi}\gamma^\mu \partial_\mu \psi \\
& - Z_4 m\overline{\psi}\psi - Z_1 e Q A_\mu \overline{\psi}\gamma^\mu \psi \\
= {}& -\frac{1}{4} F_{\mu\nu} F^{\mu\nu} - \frac{1}{2\xi} (\partial_\mu A^\mu)^2 + i\overline{\psi}\gamma^\mu \partial_\mu \psi \\
& - m\overline{\psi}\psi - e Q A_\mu \overline{\psi}\gamma^\mu \psi \\
& - \delta_3 \frac{1}{4} F_{\mu\nu} F^{\mu\nu} + \overline{\psi}(i\delta_2 \slashed{\partial} - \delta_4 m)\psi - \delta_1 e Q A_\mu \overline{\psi}\gamma^\mu \psi \qquad (7.21)
\end{aligned}
$$

To get to the last equality we separated the Z_i terms into their tree-level value 1, and the loop contributions δ_i:

$$
Z_i = 1 + \delta_i. \qquad (7.22)
$$

The terms of the Lagrangian that are proportional to δ_i are called *counterterms* and they guarantee that our theory has no more ultraviolet divergences at any order in perturbation theory. One should note that the renormalization process doesn't consist in just adding counterterms to the Lagrangian in order to subtract infinities but in a systematic order by order process which is permitted by our reinterpretation of the bare and physical quantities of the Lagrangian. After including the quantum corrections and renormalizing something else happened. The renormalized quantities are now renormalization scale (μ) dependent. Thus, one needs to solve the renormalization group equations (see Appendix B) in order to obtain their *evolution* (running) with the scale.

Our Feynman rules of the theory can now be written in terms of the renormalized quantities both in the multiplicative or the counterterm approach. For the multiplicative approach, taking a quick look at the Lagrangian one finds the following Feynman rules

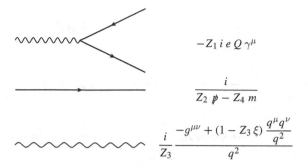

$$-Z_1\, i\, e\, Q\, \gamma^\mu$$

$$\frac{i}{Z_2\, \not{p} - Z_4\, m}$$

$$\frac{i}{Z_3}\, \frac{-g^{\mu\nu} + (1 - Z_3\, \xi)\, \dfrac{q^\mu q^\nu}{q^2}}{q^2}$$

Therefore, when summing the higher order vertex corrections or performing the Dyson resummation for the propagators (in terms of the previously introduced Feynman rules) one will directly obtain the renormalized quantities.

If one chooses the counterterm approach, the complete set of Feynman rules are given by the old ones (this time in terms of renormalized quantities)

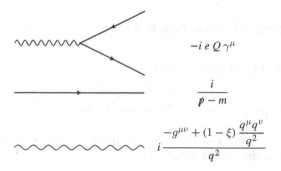

$$-i\, e\, Q\, \gamma^\mu$$

$$\frac{i}{\not{p} - m}$$

$$i\, \frac{-g^{\mu\nu} + (1 - \xi)\, \dfrac{q^\mu q^\nu}{q^2}}{q^2}$$

plus three extra ones that account for the counterterms $\sim \delta_i$:

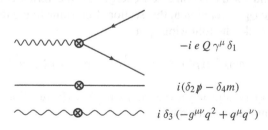

$$-i\, e\, Q\, \gamma^\mu\, \delta_1$$

$$i(\delta_2\, \not{p} - \delta_4 m)$$

$$i\, \delta_3\, (-g^{\mu\nu} q^2 + q^\mu q^\nu)$$

Thus, when calculating higher order quantum corrections, the inclusion of the counterterms will guarantee the finiteness of our result. Considering for simplicity just one-loop graphs, diagrammatically this translates into:

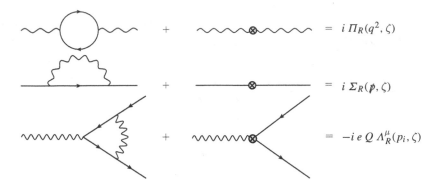

In the following, we shall study the divergent parts of the one-loop diagrams and explicitly introduce the counterterms of the theory using the two most common renormalization schemes, the $\overline{MS}$ and the on-shell (OS) scheme.

7.6 One-Loop Renormalization Example

At one-loop level the fermion self-energy is given by:

$$i\,\Sigma(\not{p}, \zeta_0) \;=\; \text{(diagram)}$$

At this order we keep corrections of $O(\alpha)$ and drop all the higher order terms, therefore we can introduce the simplified notation:

$$\Sigma(\not{p}, \zeta_0) \approx \Sigma(\not{p}, \zeta) \equiv \Sigma(\not{p}). \tag{7.23}$$

Same consideration is valid for the bosonic self-energy and the vertex correction, therefore we shall drop the ζ_0, ζ notation in this section. Returning to the fermionic self-energy, it is useful to make the following split

$$\Sigma(\not{p}) = \Sigma_1(p^2) + (\not{p} - m_0)\,\Sigma_2(p^2) \approx \Sigma_1(p^2) + (\not{p} - m)\,\Sigma_2(p^2), \tag{7.24}$$

where the last approximation is valid only at the one-loop level. The functions $\Sigma_1(p^2)$ and $\Sigma_2(p^2)$ are two Lorentz scalars given by

$$\Sigma_1(p^2) = m\,\alpha\,\frac{\mu^{2\epsilon}}{4\pi}\left(\frac{3}{\hat{\epsilon}} + \cdots\right) \equiv \Sigma_1^{\epsilon} + \Sigma_1^{R}(p^2), \tag{7.25}$$

$$\Sigma_2(p^2) = \xi\,\alpha\,\frac{\mu^{2\epsilon}}{4\pi}\left(-\frac{1}{\hat{\epsilon}} + \cdots\right) \equiv \Sigma_2^{\epsilon} + \Sigma_2^{R}(p^2), \tag{7.26}$$

where $1/\hat{\epsilon} \equiv 1/\epsilon + \gamma_E - \ln(4\pi)$ as defined in the previous chapter. The functions Σ_i^{ϵ} ($i = 1, 2$) are the parts of the self energy that contain the UV-divergent term $1/\hat{\epsilon}$ and a constant that is determined by the renormalization scheme; Σ_i^R are the regular parts of the self-energies as usual. Performing the Dyson summation for the fermionic propagator, this time including only one-loop 1PI diagrams

$$i\, S(\not{p}) =$$

we obtain:

$$S(\not{p}) = \frac{1}{\not{p} - m_0 + \Sigma(\not{p})} \equiv Z_2\, S_R(\not{p}) = \frac{Z_2}{\not{p} - m + \Sigma^R(\not{p})}. \qquad (7.27)$$

We can now in calculate Z_2 up to order α in perturbation theory. Recalling that $m = m_0 + \delta m$, is the finite renormalized mass, then:

$$
\begin{aligned}
Z_2 &= \frac{\not{p} - m + \Sigma^R}{\not{p} - m_0 + \Sigma} = \frac{(\not{p} - m)\left(1 + \Sigma_2^R\right) + \Sigma_1^R}{\not{p} - m + \delta m + \Sigma} \\
&\approx \frac{(\not{p} - m)\left(1 + \Sigma_2^R\right) + \Sigma_1^R}{\not{p} - m}\left(1 - \frac{\delta m + \Sigma}{\not{p} - m}\right) \\
&= \left(1 + \Sigma_2^R + \frac{\Sigma_1^R}{\not{p} - m}\right)\left(1 - \frac{\delta m + \Sigma_1}{\not{p} - m} - \Sigma_2\right) \\
&\approx 1 + \Sigma_2^R + \frac{\Sigma_1^R - \delta m - \Sigma_1}{\not{p} - m} - \Sigma_2 \\
&= 1 - \Sigma_2^{\epsilon} - \frac{\Sigma_1^{\epsilon} + \delta m}{\not{p} - m}. \qquad (7.28)
\end{aligned}
$$

The renormalization constants are momentum independent, thus:

$$\delta m = -\Sigma_1^{\epsilon}, \qquad\qquad Z_2 = 1 - \Sigma_2^{\epsilon}. \qquad (7.29)$$

Previously we have introduced the Z_4 constant as

$$m = m_0 + \delta m = m_0 - \Sigma_1^{\epsilon} \equiv Z_2 Z_4^{-1} m_0, \qquad (7.30)$$

therefore

$$Z_4 \approx 1 - \Sigma_2^{\epsilon} + \frac{\Sigma_1^{\epsilon}}{m_0} \approx 1 - \Sigma_2^{\epsilon} + \frac{\Sigma_1^{\epsilon}}{m}, \qquad (7.31)$$

where the last approximation is only valid at the one-loop order. We shall now move on to the analysis of the one-loop photon self energy:

$$i\,\Pi^{\mu\nu}(q^2) \; = \quad$$

As we have already mentioned, the expression of $\Pi^{\mu\nu}(q^2)$ can be factorized in the gauge invariant form:

$$\Pi^{\mu\nu}(q^2) \; = \; (-g^{\mu\nu}q^2 + q^\mu q^\nu)\,\Pi(q^2). \tag{7.32}$$

The function $\Pi(q^2)$ is given by

$$\Pi(q^2) \; = \; -\alpha\,\frac{4}{3}\,\frac{\mu^{2\epsilon}}{4\pi}\left[\frac{1}{\hat{\epsilon}} + \cdots\right] \; \equiv \; \Pi_\epsilon + \Pi_R(q^2). \tag{7.33}$$

We have decomposed as usual $\Pi(q^2)$ into an UV divergent (Π_ϵ) and a regular (Π_R) part. Performing the Dyson summation

$$i\,D^{\mu\nu}(q^2) =$$

we get:

$$\xi_0 = Z_3\,\xi, \qquad \frac{1}{1+\Pi(q^2)} = \frac{Z_3}{1+\Pi_R(q^2)}. \tag{7.34}$$

Up to order α, the Z_3 renormalization constant is given by

$$Z_3 = \frac{1+\Pi_R}{1+\Pi} \approx (1+\Pi_R)(1-\Pi_R-\Pi_\epsilon) \approx 1-\Pi_\epsilon. \tag{7.35}$$

Finally, the one-loop vertex correction is simply given by:

$$-i\,e_0\,Q\,\Lambda^\mu(p_i) \; =$$

The vertex function $\Lambda^\mu(p_i)$ explicitly reads

$$\Lambda^\mu(p_i) = \alpha\,\frac{\mu^{2\epsilon}}{4\pi}\left\{\gamma^\mu\left[-\xi\frac{1}{\hat{\epsilon}} + \cdots\right] + \cdots\right\}. \tag{7.36}$$

We can decompose as usual Λ^μ into its finite (Λ_R) and UV-divergent (Λ_ϵ) parts:

$$\Lambda^\mu(p_i) = \Lambda_\epsilon^\mu + \Lambda_R^\mu(p_i), \qquad \text{with} \qquad \Lambda_\epsilon^\mu = \gamma^\mu \Lambda_\epsilon. \tag{7.37}$$

Summing the one loop correction to the tree-level vertex one obtains

$$-i\, e_0\, Q\, \Gamma^\mu(p_i) =$$

Simplifying the $-i\, e_0\, Q$ common factor we have

$$\Gamma^\mu(p_i) = \gamma^\mu + \Lambda^\mu(p_i). \tag{7.38}$$

Introducing the Z_1 renormalization constant

$$\Gamma^\mu = \gamma^\mu(1 + \Lambda_\epsilon) + \Lambda_R^\mu = \frac{1}{Z_1}\Gamma_R^\mu = \frac{1}{Z_1}(\gamma^\mu + \Lambda_R^\mu), \tag{7.39}$$

up to order α we obtain

$$Z_1\left(\gamma^\mu(1 + \Lambda_\epsilon) + \Lambda_R^\mu\right) \approx Z_1\gamma^\mu(1 + \Lambda_\epsilon) + \Lambda_R^\mu = \gamma^\mu + \Lambda_R^\mu. \tag{7.40}$$

Thus

$$Z_1 = 1 - \Lambda_\epsilon. \tag{7.41}$$

In conclusion, the one-loop level QED renormalization constants are given by:

$Z_1 = 1 - \Lambda_\epsilon$	$\Rightarrow$	$\delta_1 = -\Lambda_\epsilon$
$Z_3 = 1 - \Pi_\epsilon$	$\Rightarrow$	$\delta_3 = -\Pi_\epsilon$
$Z_2 = 1 - \Sigma_2^\epsilon$	$\Rightarrow$	$\delta_2 = -\Sigma_2^\epsilon$
$Z_4 = 1 - \Sigma_2^\epsilon + \Sigma_1^\epsilon/m$	$\Rightarrow$	$\delta_4 = -\Sigma_2^\epsilon + \Sigma_1^\epsilon/m$	(7.42)

with $Z_1 = Z_2$ therefore, $\Sigma_2^\epsilon = \Lambda_\epsilon$, as it can be explicitly checked.

It is worth making the following comment. As the photon always couples to conserved currents, the $q^\mu q^\nu$ part of the propagator will always vanish when calculating physical observables. Thus, one can simplify the renormalization procedure by choosing for example the Feynman gauge ($\xi = 1$) for the tree-level photon propagator and, when summing higher order corrections, simply ignore the $q^\mu q^\nu$

terms. However, in this chapter I preferred giving the full formal renormalization prescription.

7.6.1 $\overline{MS}$ Renormalization Scheme

The split we have just made between the UV divergent and the regular parts of the previous functions is somewhat ambiguous. A given choice of this separation defines a *renormalization scheme*. For the $\overline{MS}$ scheme, the Z_i renormalization constants only absorb the $1/\hat{\epsilon}$ terms (plus no other constant) thus, they are simply given by:

$$Z_2 = Z_1 = 1 + \frac{\alpha}{4\pi} \mu^{2\epsilon} \xi \frac{1}{\hat{\epsilon}}, \tag{7.43}$$

$$Z_3 = 1 + \frac{\alpha}{3\pi} \mu^{2\epsilon} \frac{1}{\hat{\epsilon}}, \tag{7.44}$$

$$Z_4 = 1 + \frac{\alpha}{4\pi} \mu^{2\epsilon} (3 + \xi) \frac{1}{\hat{\epsilon}}. \tag{7.45}$$

Next we present the on-shell (OS) renormalization scheme.

7.6.2 On-Shell Renormalization Scheme

For this renormalization scheme the expressions of the counterterms are not as straightforward to obtain as previously. In the OS scheme the renormalization counterterms must guarantee that the the poles of the renormalized propagators coincide with the physical masses (m for the fermion and 0 for the photon in our case). The standard procedure is shown next.

Let's take a look at the fermionic propagator (7.27) and make a Taylor expansion around the pole ($\not{p} = m$)

$$\frac{1}{\not{p} - m_0 + \Sigma(\not{p})} \approx \frac{1}{\not{p} - m_0 + \Sigma(m) + (\not{p} - m)\Sigma'(m)}$$

$$\approx \frac{Z_2}{\not{p} - m + \Sigma^R(m) + (\not{p} - m)\Sigma'^R(m)}, \tag{7.46}$$

where $\Sigma'(m)$ is the short-hand notation for

$$\Sigma'(m) = \frac{d}{d\not{p}} \Sigma(\not{p})\Big|_{\not{p}=m} \tag{7.47}$$

(when differentiating with respect to p one must take into account that $p^2 = p̸^2$, thus $dp^2/dp̸ = 2p̸$). Imposing that $\Sigma^R(m) = 0$ and $\Sigma'^R(m) = 0$ (the pole of the renormalized propagator is given by the physical mass with same residue 1 as the tree-level propagator) we have

$$\frac{1}{p̸ - m + \delta m + \Sigma(m) + (p̸ - m)\Sigma'(m)} \approx \frac{Z_2}{p̸ - m}. \tag{7.48}$$

Choosing the mass counterterm to be

$$\delta m = -\Sigma(m), \tag{7.49}$$

we obtain the Z_2 renormalization constant at one-loop level

$$Z_2 = 1 - \Sigma'(m) = Z_1. \tag{7.50}$$

Thus, after having defined the previous renormalization constants, looking again at (7.27) we find that for the on-shell scheme, at the one-loop level, the regular part of the self-energy is given by

$$\Sigma^R(p̸) = \Sigma(p̸) - \Sigma(m) - (p̸ - m)\,\Sigma'(m). \tag{7.51}$$

Looking at (7.34), for $q^2 = 0$ we have the following relation

$$\frac{1}{1 + \Pi(0)} = \frac{Z_3}{1 + \Pi_R(0)}. \tag{7.52}$$

Imposing the on-shell condition $\Pi_R(0) = 0$ we obtain the Z_3 renormalization constant:

$$Z_3 = 1 - \Pi(0). \tag{7.53}$$

As in the previous case, looking again at (7.34), for arbitrary q^2, the regular part of the bosonic self-energy reads:

$$\Pi_R(q^2) = \Pi(q^2) - \Pi(0). \tag{7.54}$$

Any other condition imposed on the vertex would be redundant. As we have already mentioned many times, gauge invariance fixes $Z_1 = Z_2$, and therefore the value of the regular part of vertex function is also fixed by the previous on-shell conditions.

So far we have supposed that the fermion of the theory is a stable particle i.e., it does not decay into lighter particles. If the fermion can however, decay (a muon for example), we must slightly modify the previous on-shell conditions for the fermion mass and wave function renormalization (besides of course, including and renormalizing the new terms of the Lagrangian). If $p^2 \geq \left(\sum_i m_i\right)^2$, where m_i are the masses of the particles in which the fermion can decay, then $\Sigma(p)$ develops an imaginary part that is UV-finite (thus, it needs no renormalization). Separating the self energy into the real and imaginary parts, (7.27) takes the form:

$$\frac{1}{\not{p} - m_0 + \mathrm{Re}\,\Sigma(\not{p}) + i\,\mathrm{Im}\,\Sigma(\not{p})} = \frac{Z_2}{\not{p} - m + \mathrm{Re}\,\Sigma^R(\not{p}) + i\,\mathrm{Im}\,\Sigma(\not{p})}. \quad (7.55)$$

At one-loop order:

$$\not{p} - m_0 + \mathrm{Re}\,\Sigma(\not{p}) + i\,\mathrm{Im}\,\Sigma(\not{p}) = Z_2^{-1}\left(\not{p} - m + \mathrm{Re}\,\Sigma^R(\not{p}) + i\,\mathrm{Im}\,\Sigma(\not{p})\right)$$
$$\approx (1 - \delta_2)(\not{p} - m) + \mathrm{Re}\,\Sigma^R(\not{p}) + i\,\mathrm{Im}\,\Sigma(\not{p}). \quad (7.56)$$

Thus, at this order in perturbation theory

$$\frac{1}{\not{p} - m_0 + \mathrm{Re}\,\Sigma(\not{p})} \approx \frac{Z_2}{\not{p} - m + \mathrm{Re}\,\Sigma^R(\not{p})}. \quad (7.57)$$

Using the same procedure as previously (making a Taylor expansion around the pole, this time only for the real part of the self-energy) we obtain

$$\delta m = -\mathrm{Re}\,\Sigma(m), \qquad Z_2 = 1 - \mathrm{Re}\,\Sigma'(m) = Z_1. \quad (7.58)$$

and

$$\Sigma^R(\not{p}) = \mathrm{Re}\,\Sigma(\not{p}) - \mathrm{Re}\,\Sigma(m) - (\not{p} - m)\,\mathrm{Re}\,\Sigma'(m) + i\,\mathrm{Im}\,\Sigma(\not{p}). \quad (7.59)$$

The imaginary part of the self-energy shifts the pole of the propagator from the real to the imaginary axis. When reaching the on-shell region ($p^2 = m^2$), this imaginary part takes the well known Breit-Wigner form

$$i\,\mathrm{Im}\,\Sigma(m) = i\,m\,\Gamma, \quad (7.60)$$

where Γ is the total decay rate of the fermion (see Chap. 12 for details and demonstration). This has a physical consequence. It is obviously the reason why, experimentally, a resonance doesn't peak as a delta function, but has a finite width.

As the Imaginary part of the self-energy does not affect the UV-divergent poles, the $\overline{MS}$-scheme must not be modified for unstable particles.

7.7 Renormalization and Tadpoles

Depending on the quantization procedure one might have to deal with tadpoles. This topology is shown in Fig.7.1. The contribution of this diagram modifies the vacuum expectation value (vev) of the field as we shall see in a moment. If one uses the canonical quantization procedure with *normal ordered* products of fields one will never find this topology as it involves the contraction of two fields in the same space-time point. If one uses the path integral quantization, one finds the same topologies for the connected diagrams as in the previous case plus tadpoles. Thus, when dealing with propagator corrections one will find topologies like

which are not proper 1PI diagrams. Do not confuse tadpole topologies with the *seagull* topologies that are shown below

The seagull topologies (which are oftenly also called tadpoles by many authors) are proper 1PI diagrams and **must** be included into the renormalization algorithm.

Returning to tadpoles, in order to get rid of these contributions one usually redefines the field's vev and generates a counter-term in the Lagrangian that cancels the tadpoles to all orders in perturbation theory. We can explicitly see this with the following example. Consider the ϕ^3 Lagrangian

$$\mathcal{L}(x) = \frac{1}{2}\partial^\mu \phi(x)\partial_\mu \phi(x) - \frac{1}{2}m^2\phi^2(x) - \frac{\lambda}{3!}\phi^3(x). \tag{7.61}$$

Including one-loop order corrections one finds that the vev of ϕ is given by

$$\langle 0|\phi(x)|0\rangle = \frac{t}{m^2}. \tag{7.62}$$

Fig. 7.1 Tadpole topology

$$it = $$

with t shown in Fig. 7.1. Defining a new field $\phi'(x) = \phi(x) - t/m^2$, one immediately finds

$$\langle 0|\phi(x)'|0\rangle = \langle 0|\phi(x)|0\rangle - \frac{t}{m^2}\langle 0|0\rangle = 0. \tag{7.63}$$

Rewriting the Lagrangian (only the parts that correspond to the free field, obviously the interaction term must not be modified as it was already used to calculate t) in terms of the field $\phi'(x)$ one generates a counterterm of the type $\mathcal{L}_{ct}(x) = -\phi'(x)\,t$ which cancels the tadpole contributions.[1] Diagrammatically

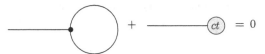

for the vev and

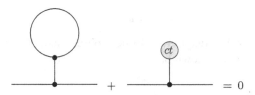

for the propagator correction. So, in this simple ϕ^3 model one does not even need to bother in calculating tadpoles. They simply cancel when the counterterms are added. Thus, one can consider as usual, just 1PI diagrams when renormalizing the Lagrangian. Defining the Z_i renormalization constants as

$$m_0^2 = Z_m\,m^2 \qquad \phi_0 = Z_\phi^{1/2}\,\phi, \qquad \lambda_0 = Z_\lambda\,\lambda, \tag{7.64}$$

and dropping the *primed* notation one finally finds the following Lagrangian in terms of the renormalized quantities

$$\mathcal{L} = Z_\phi\frac{1}{2}\partial^\mu\phi\,\partial_\mu\phi - Z_m\,Z_\phi\frac{1}{2}m^2\,\phi^2 - Z_\lambda\,Z_\phi^{3/2}\frac{\lambda}{3!}\phi^3 - t\,\phi. \tag{7.65}$$

This is however, a toy model. Things are not that simple in the electroweak sector of the Standard Model. The Higgs field will generate tadpoles. After performing a similar redefinition one will generate a similar counterterm. However, there will be terms in the Lagrangian that will generate diagrams of the type

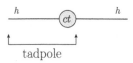
tadpole

---

[1]One also generates a constant term $-t^2/2m^2$ that can be safely dropped as it will play no role in the interactions.

which must be included in the renormalization process (they also play an important role in the Ward identities). It is also worth mentioning that tadpoles also play an important role is one wishes to define gauge independent mass counterterms,[2] which is not mandatory (mass counterterms are not observables) but it is a nice feature that one can exploit. More on tadpoles and the Ward identities in the Standard Model will be treated in Chap. 9.

Finally, tadpoles for vector fields (massive or massless) in the Standard Model always vanish, as they would simply break Lorentz invariance (which is one of the fundamental symmetries of the theory).

Further Reading

M. Kaku, *Quantum Field Theory: A Modern Introduction* (Oxford University Press, New York, 1993)

A. Pich, *Class Notes on Quantum Field Theory.* http://eeemaster.uv.es/course/view.php?id=6

M. Srednicki, *Quantum Field Theory* (Cambridge University Press, New York)

L.H. Ryder, *Quantum Field Theory* (Cambridge University Press, New York, 1985)

T.P. Cheng, L.F. Li, *Gauge Theory of Elementary Particle Physics* (Oxford University Press, New York, 1984)

F. Mandl, G.P. Shaw, *Quantum Field Theory* (Chichester, New York, 1984)

---

[2]The full details at one and two-loop level are very nicely given in S. Actis, A. Ferroglia, M. Passera and G. Passarino, *Two-Loop Renormalization in the Standard Model. Part I: Prolegomena*, Nucl. Phys. B **777** (2007) 1, http://arxiv.org/pdf/hep-ph/0612122.pdf

Chapter 8
One-Loop Two and Three-Point Functions

Abstract In this chapter we present a few relevant calculations of one-loop, one and two-point (scalar, vector and tensor) functions. IR and UV divergences are extensively treated. One example of IR-pole cancellation is presented. The two and three-body phase space integrals in D dimensions, needed for the calculation of IR divergent cross sections are also given. Last, the usage of the generic parametrization (6.27) for non-integer powers of propagators (which appear when one needs to integrate over the four-momentum, logarithmic functions that depend on the four-momentum) is shown with a simple two-loop example. With the tools given here, the reader should find straightforward to construct any higher order scalar or tensor integral for any N-point function at one-loop level.

8.1 Two-Point Function

We shall begin this chapter by studying the two-point function. The reader is highly encouraged to perform all the calculations step-by-step and reproduce the results obtained here.

The distribution of momenta is shown in Fig. 8.1. The exterior legs are not necessarily on-shell (meaning that we shall not use the equality $p_i^2 = m_i^2$). All the masses are considered to be different. The denominator of this two-point function is

$$\frac{1}{[(k + p_1)^2 - m^2](k^2 - M^2)}. \tag{8.1}$$

The Feynman parametrization that we will use in this case is (6.23). Taking $A = [(k + p_1)^2 - m^2]$ and $B = (k^2 - M^2)$ we obtain:

$$\frac{1}{[(k + p_1)^2 - m^2](k^2 - M^2)} = \int_0^1 dx \frac{1}{\mathcal{D}^2}. \tag{8.2}$$

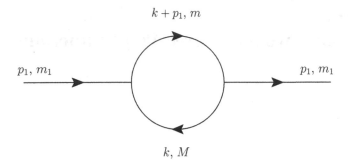

Fig. 8.1 Distribution of masses and momenta for the two-point function

where $\mathcal{D}$ has the following expression:

$$\mathcal{D} = (k + p_1 x)^2 - a^2, \qquad a^2 \equiv p_1^2 x(x-1) + m^2 x + M^2(1-x). \qquad (8.3)$$

The first integral that we shall deal with is the scalar integral:

$$\boxed{I_1 \equiv \int \frac{d^D k}{(2\pi)^D} \frac{1}{[(k+p_1)^2 - m^2](k^2 - M^2)}.} \qquad (8.4)$$

Using the previous results for dimensional regularization from Chap. 6 we find

$$
\begin{aligned}
I_1 &= \int_0^1 dx \int \frac{d^D k}{(2\pi)^D} \frac{1}{[(k+p_1 x)^2 - a^2]^2} \\
&= \int_0^1 dx \int \frac{d^D k}{(2\pi)^D} \frac{1}{[k^2 - a^2]^2} \\
&= \int_0^1 dx\, J(D, 0, 2, a^2) \\
&= \frac{-i}{(4\pi)^2} \mu^{2\epsilon} \left[\frac{1}{\hat{\epsilon}} + \int_0^1 dx\, \ln\left(\frac{a^2}{\mu^2}\right) \right],
\end{aligned}
\qquad (8.5)
$$

where, to get to the second line we shifted the integration variable $k \to k - x\, p_1$. The second integral that we shall calculate is the vector integral:

$$\boxed{I_2^\mu \equiv \int \frac{d^D k}{(2\pi)^D} \frac{k^\mu}{[(k+p_1)^2 - m^2](k^2 - M^2)}.} \qquad (8.6)$$

This is exactly the one we have studied in Chap. 6:

$$
\begin{aligned}
I_2^{\mu} &= \int_0^1 dx \int \frac{d^D k}{(2\pi)^D} \frac{k^{\mu}}{[(k+p_1 x)^2 - a^2]^2} \\
&= p_1^{\mu} \frac{i}{(4\pi)^2} \mu^{2\epsilon} \left[\frac{1}{2\hat{\epsilon}} + \int_0^1 dx\, x \, \ln\left(\frac{a^2}{\mu^2}\right) \right].
\end{aligned}
\tag{8.7}
$$

The third integral we analyse is another scalar integral:

$$
\boxed{ I_3 \equiv \int \frac{d^D k}{(2\pi)^D} \frac{k^2}{[(k+p_1)^2 - m^2](k^2 - M^2)}. }
\tag{8.8}
$$

The calculation is a little more complicated but equally straightforward:

$$
\begin{aligned}
I_3 &= \int_0^1 dx \int \frac{d^D k}{(2\pi)^D} \frac{k^2}{[(k+p_1 x)^2 - a^2]^2} \\
&= \int_0^1 dx \int \frac{d^D k}{(2\pi)^D} \frac{(k - x p_1)^2}{[k^2 - a^2]^2} \\
&= \int_0^1 dx \left[J(D, 1, 2, a^2) + p_1^2 x^2 J(D, 0, 2, a^2) \right] \\
&= \int_0^1 dx \left(\frac{a^2 D}{D-2} + p_1^2 x^2 \right) J(D, 0, 2, a^2) \\
&= \int_0^1 dx \left(a^2(2 - \epsilon) + p_1^2 x^2 \right) J(D, 0, 2, a^2) \\
&= \frac{-i}{(4\pi)^2} \mu^{2\epsilon} \left[\frac{1}{\hat{\epsilon}}(m^2 + M^2) - \frac{1}{2}\left(m^2 + M^2 - \frac{1}{3} p_1^2\right) \right. \\
&\qquad\qquad \left. + \int_0^1 dx (p_1^2 x^2 + 2a^2) \ln\left(\frac{a^2}{\mu^2}\right) \right].
\end{aligned}
\tag{8.9}
$$

As usual we have ignored terms of $O(\epsilon)$. The next interesting case is the following tensor integral

$$
\boxed{ I_4^{\mu\nu} \equiv \int \frac{d^D k}{(2\pi)^D} \frac{k^{\mu} k^{\nu}}{[(k+p_1)^2 - m^2](k^2 - M^2)}. }
\tag{8.10}
$$

We find the following expression for $I_4^{\mu\nu}$:

$$
\begin{aligned}
I_4^{\mu\nu} &= \int_0^1 dx \int \frac{d^D k}{(2\pi)^D} \frac{k^\mu k^\nu}{[(k+p_1 x)^2 - a^2]^2} \\
&= \int_0^1 dx \int \frac{d^D k}{(2\pi)^D} \frac{(k^\mu - x p_1^\mu)(k^\nu - x p_1^\nu)}{[k^2 - a^2]^2} \\
&= \int_0^1 dx \left(\frac{g^{\mu\nu}}{D} \frac{a^2 D}{D-2} + p_1^\mu p_1^\nu x^2 \right) J(D,0,2,a^2) \\
&= \frac{-i}{(4\pi)^2} \mu^{2\epsilon} \left[\left(\frac{g^{\mu\nu}}{4}(m^2 + M^2 - \tfrac{1}{3}p_1^2) + \tfrac{1}{3}p_1^\mu p_1^\nu \right) \frac{1}{\epsilon} \right. \\
&\qquad\qquad - \frac{g^{\mu\nu}}{4}\left(m^2 + M^2 - \tfrac{1}{3}p_1^2\right) \\
&\qquad\qquad \left. + \int_0^1 dx \left(\frac{g^{\mu\nu}}{2}a^2 + p_1^\mu p_1^\nu x^2 \right) \ln\left(\frac{a^2}{\mu^2} \right) \right].
\end{aligned}
\tag{8.11}
$$

The next integral we study is again a scalar type integral:

$$
I_5 \equiv \int \frac{d^D k}{(2\pi)^D} \frac{(k^2)^2}{[(k+p_1)^2 - m^2](k^2 - M^2)}.
\tag{8.12}
$$

We obtain:

$$
\begin{aligned}
I_5 &= \int_0^1 dx \int \frac{d^D k}{(2\pi)^D} \frac{(k^2)^2}{[(k+p_1 x)^2 - a^2]^2} \\
&= \int_0^1 dx \int \frac{d^D k}{(2\pi)^D} \frac{2x^2 p_1^2 k^2 + \frac{4}{D}x^2 p_1^2 k^2 + (k^2)^2 + x^4(p_1^2)^2}{[k^2 - a^2]^2} \\
&= \int_0^1 dx \left[\left(2x^2 p_1^2 + \frac{4}{D}x^2 p_1^2\right) \frac{a^2 D}{D-2} + \frac{a^4(D+2)}{D-2} + x^4(p_1^2)^2 \right] J(D,0,2,a^2) \\
&= \frac{-i}{(4\pi)^2}\mu^{2\epsilon} \left[\frac{1}{\epsilon}\left(m^2(p_1^2 + M^2) + m^4 + M^4 \right) \right. \\
&\qquad\qquad + \frac{2}{3}\left(\frac{(p_1^2)^2}{5} - m^2(p_1^2 + M^2) - m^4 - M^4 \right) \\
&\qquad\qquad \left. + \int_0^1 dx \left(3a^4 + 6a^2 x^2 p_1^2 + x^4(p_1^2)^2 \right) \ln\left(\frac{a^2}{\mu^2} \right) \right].
\end{aligned}
\tag{8.13}
$$

The last integral that we are going to analyse is

$$
I_6^\mu \equiv \int \frac{d^D k}{(2\pi)^D} \frac{k^\mu k^2}{[(k+p_1)^2 - m^2](k^2 - M^2)}.
\tag{8.14}
$$

Fig. 8.2 One-loop fermionic
two-point function in QED

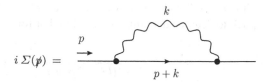

$$i\,\Sigma(\not{p}) =$$

We obtain the following:

$$
I_6^\mu = \int_0^1 dx \int \frac{d^D k}{(2\pi)^D} \frac{k^\mu k^2}{[(k+p_1 x)^2 - a^2]^2}
$$

$$
= -p_1^\mu \int_0^1 dx \left[x\left(1 + \frac{2}{D}\right)\frac{a^2 D}{D-2} + x^3 p_1^2 \right] J(D,0,2,a^2)
$$

$$
= p_1^\mu \frac{i}{(4\pi)^2}\mu^{2\epsilon} \int_0^1 dx \left[\frac{1}{\epsilon}\left(3a^2 x + x^3 p_1^2\right) - 2a^2 x + \left(3a^2 x + x^3 p_1^2\right)\ln\left(\frac{a^2}{\mu^2}\right) \right]
$$

$$
= p_1^\mu \frac{i}{(4\pi)^2}\mu^{2\epsilon}\left[\frac{1}{\epsilon}\left(m^2 + \frac{M^2}{2}\right) + \frac{1}{3}\left(\frac{p_1^2}{2} - 2m^2 - M^2\right) \right.
$$

$$
\left. + \int_0^1 dx \left(3a^2 x + x^3 p_1^2\right)\ln\left(\frac{a^2}{\mu^2}\right) \right]. \tag{8.15}
$$

In the same fashion, one should find straightforward to calculate any one-loop integral for any arbitrary tensor function in the numerator. Next we will analyse a two-point function that presents both IR and UV poles.

8.2 IR Divergences and the Two-Point Function

Besides the ultraviolet divergences we have seen until now, which are treated consistently within the renormalization procedure there are also other types of divergences called infrared, as we have seen in Chap. 6. Consider the one-loop fermionic QED two-point function for massless on-shell fermions in the Feynman gauge $\xi = 1$ (Fig. 8.2). It reads[1]:

$$
\boxed{ i\,\Sigma(\not{p}) = -e^2 \int \frac{d^D k}{(2\pi)^D} \frac{\gamma^\mu(\not{p}+\not{k})\gamma_\mu}{k^2(k+p)^2} }, \tag{8.16}
$$

---

[1] The QED Feynman rules are given in Chap. 5.

with $p^2 \to 0$ and where we have taken $Q^2 = 1$ for the electric charge. Taking a quick look at the denominator we find that for small k

$$\frac{1}{k^2(k^2 + 2p \cdot k)} \xrightarrow{k^2 \ll} \frac{1}{k^2(2p \cdot k)}. \tag{8.17}$$

We know that denominator diverges in two cases:

$$\begin{array}{lll} 1. \ k \to 0, & \text{soft divergence,} & \\ 2. \ p \cdot k \to 0, & \text{collinear divergence.} & \end{array} \tag{8.18}$$

It is worth mentioning that a typical way of regulating these divergences is by giving small masses m_γ, m_e both to the photon and the electron and taking the limit $m_\gamma, m_e \to 0$ at the end. However, the most elegant way is still, *Dimensional Regularization* (doesn't break the symmetries of our theory). As seen in Chap. 6, we must take $D = 4 + 2\epsilon'$, with $\epsilon' > 0$ for the IR divergent parts and $D = 4 + 2\epsilon$, with $\epsilon < 0$ for the UV divergent parts. In D dimensions the two-point function (8.16) reads:

$$\begin{aligned} i\Sigma(p) &= -e^2 \int \frac{d^D k}{(2\pi)^D} \frac{(2 - D)(\not{p} + \not{k})}{k^2(k + p)^2} \\ &= -e^2 \int_0^1 dx \int \frac{d^D k}{(2\pi)^D} \frac{(2 - D)(\not{p} + \not{k})}{[(k + xp)^2 - a^2]^2} \\ &= -e^2(2 - D)\not{p} \int_0^1 dx(1 - x)J(D, 0, 2, a^2). \end{aligned} \tag{8.19}$$

where we have defined $a^2 \equiv -p^2 x(1 - x)$. Now, instead of writing $J(D, 0, 2, a^2)$ as (6.9), let's write it in the form of (6.2):

$$\begin{aligned} i\Sigma(p) &= \frac{-i}{(4\pi)^{D/2}} \not{p} e^2(2 - D)\Gamma(2 - D/2) \int_0^1 dx(1 - x)(a^2)^{D/2 - 2} \\ &= \frac{-i}{(4\pi)^{D/2}} \not{p} e^2(2 - D)\Gamma(2 - D/2)(-p^2)^{D/2 - 2} \\ &\quad \times \int_0^1 dx(1 - x)^{D/2 - 1} x^{D/2 - 2}. \end{aligned} \tag{8.20}$$

Using the Euler Beta function (A.17) we find the following:

$$i\Sigma(p) = \frac{-i}{(4\pi)^{D/2}} \not{p} e^2(2 - D)\frac{\Gamma(2 - D/2)\Gamma(D/2)\Gamma(D/2 - 1)}{\Gamma(D - 1)}(-p^2)^{D/2 - 2}. \tag{8.21}$$

We have an UV pole at $D = 4$ coming from $\Gamma(2 - D/2)$ and we have an indetermination of the type 0^0 for $(-p^2)^{D/2 - 2}$ when $p^2 \to 0$ and $D \to 4$, which will appear in our equations in the form of IR and UV poles. In order to regularize this, we must use the following integral:

$$\boxed{\int_{-y}^{\infty} \frac{dx}{x} x^{\epsilon} = (-y)^{\epsilon} \left(-\frac{1}{\epsilon}\right)}. \tag{8.22}$$

Therefore, $(-p^2)^{D/2-2}$ can be written in the following form

$$
\begin{aligned}
(-p^2)^{D/2-2} &= (2 - D/2) \int_{-p^2}^{\infty} \frac{dq^2}{q^2} (q^2)^{D/2-2} \\
&= (2 - D/2)(\mu^2)^{D/2-2} \int_{-p^2}^{\infty} \frac{dq^2}{q^2} \left(\frac{q^2}{\mu^2}\right)^{D/2-2} \\
&= (2 - D/2)(\mu^2)^{D/2-2} \Bigg[\int_{-p^2}^{\mu^2} \frac{dq^2}{q^2} \left(\frac{q^2}{\mu^2}\right)^{D/2-2} \\
&\quad + \int_{\mu^2}^{\infty} \frac{dq^2}{q^2} \left(\frac{q^2}{\mu^2}\right)^{D/2-2} \Bigg]. \tag{8.23}
\end{aligned}
$$

In order to make these integrations we need the following intermediate results

$$\boxed{\begin{aligned}
\int_{0}^{\mu} dx \frac{1}{x} \left(\frac{x}{\mu}\right)^{\epsilon} &= \frac{1}{\epsilon} \text{ for } \mathrm{Re}(\epsilon) > 0 \text{ and } \mu > 0 \\
\int_{\mu}^{\infty} dx \frac{1}{x} \left(\frac{x}{\mu}\right)^{\epsilon} &= -\frac{1}{\epsilon} \text{ for } \mathrm{Re}(\epsilon) < 0 \text{ and } \mu > 0
\end{aligned}} \tag{8.24}$$

Therefore, we can safely take $p^2 \to 0$ in (8.23) and we obtain

$$(-p^2)^{D/2-2} = (2 - D/2)(\mu^2)^{D/2-2} \left[\frac{1}{\epsilon'} - \frac{1}{\epsilon}\right]. \tag{8.25}$$

Taking D adequately in each case we find

$$i\Sigma(\not{p}) = \frac{-i}{(4\pi)^2} \not{p} e^2 \left[-\left(\frac{\mu^2}{4\pi}\right)^{\epsilon'} \left(\frac{1}{\epsilon'} + \gamma_E\right) + \left(\frac{\mu^2}{4\pi}\right)^{\epsilon} \left(\frac{1}{\epsilon} + \gamma_E\right) \right], \tag{8.26}$$

where we have summed and subtracted γ_E and we have ignored all terms of $O(\epsilon)$ and $O(\epsilon')$. Introducing our standard notation for the UV pole (6.7) and making a similar definition for the IR pole

$$\boxed{\frac{1}{\tilde{\epsilon}'} \equiv \frac{1}{\epsilon'} + \gamma_E - \ln(4\pi)}, \tag{8.27}$$

we finally obtain

$$\boxed{i\,\Sigma(\not{p}) = -i\,\not{p}\frac{\alpha}{4\pi}\left[-\mu^{2\epsilon'}\left(\frac{1}{\hat{\epsilon}'}\right)+\mu^{2\epsilon}\left(\frac{1}{\hat{\epsilon}}\right)\right].}$$
(8.28)

The usual renormalization approach is to also absorb the IR pole into the renormalization constant. Therefore we have $Z_1 = Z_2 = 1 + \delta_1 = 1 + \delta_2$ (remember $Z_1 = Z_2$, see Chap. 7) with

$$\delta_1 = \delta_2 = \frac{\alpha}{4\pi}\left[-\mu^{2\epsilon'}\left(\frac{1}{\hat{\epsilon}'}\right)+\mu^{2\epsilon}\left(\frac{1}{\hat{\epsilon}}\right)\right],$$
(8.29)

for the $\overline{MS}$-scheme. Note that the renormalized self-energy $\Sigma_R(\not{p}) = 0$ in this case. One should also realize that Z_1 and Z_2 do not contribute to the renormalization of any physical quantity like the electric charge or the mass and therefore, the running (with the renormalization scale) of the these physical quantities is not affected by the IR pole we have included in $\delta_{1,2}$ (as it must, the running of physical quantities only depends on the UV-poles and never on the IR-poles). This last result, (8.29) will turn out to be extremely useful within a few sections, where we will analyse an example of IR-pole cancellation.

Next, we shall move on and analyse the three-point functions. The IR divergences that will appear will be much easier to localize and to treat.

8.3 Three-Point Function

Momentum conservation in this case reads

$$p_1 = p_2 + p_3,$$
(8.30)

and as usual, the exterior legs are not necessarily on-shell. The arrows in Fig. 8.3 indicate the distribution of momenta, and all the masses are considered to be different. The denominator of this three-point function is:

$$\frac{1}{(k^2 - M^2)[(k + p_2)^2 - m^2][(k - p_3)^2 - \overline{m}^2]}.$$
(8.31)

The Feynman parametrization that we shall use in this case is (6.25). Taking $A = (k^2 - M^2)$, $B = [(k + p_2)^2 - m^2]$ and $C = [(k - p_3)^2 - \overline{m}^2]$ we obtain:

$$\frac{1}{(k^2 - M^2)[(k + p_2)^2 - m^2][(k - p_3)^2 - \overline{m}^2]} = \int_0^1 dy \int_0^{1-y} dz \frac{2}{D^3}.$$
(8.32)

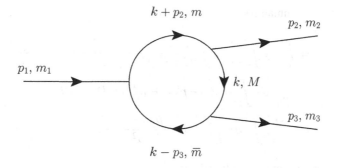

Fig. 8.3 Distribution of masses and momenta of the three-point function

with $\mathcal{D}$ given by:

$$\mathcal{D} = (k + p_2 y - p_3 z)^2 - b^2, \tag{8.33}$$

$$b^2 = m^2 y + \overline{m}^2 z + M^2 (1 - y - z) + p_3^2 z(z - 1)$$
$$+ p_2^2 y(y - 1) - 2yz(p_2 \cdot p_3). \tag{8.34}$$

The first integral that we shall deal with is the scalar integral:

$$\boxed{C_1 \equiv \int \frac{d^D k}{(2\pi)^D} \frac{1}{(k^2 - M^2)[(k + p_2)^2 - m^2][(k - p_3)^2 - \overline{m}^2]}.} \tag{8.35}$$

Introducing the Feynman parametrization and manipulating the expression:

$$C_1 = \int_0^1 dy \int_0^{1-y} dz \int \frac{d^D k}{(2\pi)^D} \frac{2}{[(k + p_2 y - p_3 z)^2 - b^2]^3}$$
$$= \int_0^1 dy \int_0^{1-y} dz \int \frac{d^D k}{(2\pi)^D} \frac{2}{[k^2 - b^2]^3}$$
$$= 2 \int_0^1 dy \int_0^{1-y} dz J(D, 0, 3, b^2)$$
$$= \frac{-i}{(4\pi)^2} \int_0^1 dy \int_0^{1-y} dz \frac{1}{b^2}. \tag{8.36}$$

The second integral we wish to calculate is the vector integral:

$$\boxed{C_2^\mu \equiv \int \frac{d^D k}{(2\pi)^D} \frac{k^\mu}{(k^2 - M^2)[(k + p_2)^2 - m^2][(k - p_3)^2 - \overline{m}^2]}.} \tag{8.37}$$

Introducing the Feynman parametrization we obtain:

$$
\begin{aligned}
C_2^\mu &= \int_0^1 dy \int_0^{1-y} dz \int \frac{d^D k}{(2\pi)^D} \frac{2k^\mu}{[(k + p_2 y - p_3 z)^2 - b^2]^3} \\
&= 2 \int_0^1 dy \int_0^{1-y} dz (-p_2^\mu y + p_3^\mu z) J(D, 0, 3, b^2) \\
&= \frac{-i}{(4\pi)^2} \int_0^1 dy \int_0^{1-y} dz \frac{-p_2^\mu y + p_3^\mu z}{b^2}.
\end{aligned}
$$
(8.38)

The next integral we are going to calculate is:

$$
\boxed{C_3 \equiv \int \frac{d^D k}{(2\pi)^D} \frac{k^2}{(k^2 - M^2)[(k + p_2)^2 - m^2][(k - p_3)^2 - \overline{m}^2]}.}
$$
(8.39)

Manipulating the expression we get:

$$
\begin{aligned}
C_3 &= \int_0^1 dy \int_0^{1-y} dz \int \frac{d^D k}{(2\pi)^D} \frac{2k^2}{[(k + p_2 y - p_3 z)^2 - b^2]^3} \\
&= 2 \int_0^1 dy \int_0^{1-y} dz \left[\left(1 + \frac{\epsilon}{2}\right) J(D, 0, 2, b^2) + (p_2 y - p_3 z)^2 J(D, 0, 3, b^2) \right] \\
&= \frac{-i}{(4\pi)^2} \mu^{2\epsilon} \int_0^1 dy \int_0^{1-y} dz \left[\frac{2}{\hat{\epsilon}} + 2\ln\left(\frac{b^2}{\mu^2}\right) + 1 + \frac{1}{b^2}(p_2 y - p_3 z)^2 \right] \\
&= \frac{-i}{(4\pi)^2} \mu^{2\epsilon} \left\{ \frac{1}{\hat{\epsilon}} + \frac{1}{2} + \int_0^1 dy \int_0^{1-y} dz \left[\frac{(p_2 y - p_3 z)^2}{b^2} + 2\ln\left(\frac{b^2}{\mu^2}\right) \right] \right\}.
\end{aligned}
$$
(8.40)

The last integral that we will analyse in this section is tensor integral $C_4^{\mu\nu}$:

$$
\boxed{C_4^{\mu\nu} \equiv \int \frac{d^D k}{(2\pi)^D} \frac{k^\mu k^\nu}{(k^2 - M^2)[(k + p_2)^2 - m^2][(k - p_3)^2 - \overline{m}^2]}.}
$$
(8.41)

We obtain the following result:

$$
\begin{aligned}
C_4^{\mu\nu} &= \int_0^1 dy \int_0^{1-y} dz \int \frac{d^D k}{(2\pi)^D} \frac{2k^\mu k^\nu}{[(k + p_2 y - p_3 z)^2 - b^2]^3} \\
&= 2 \int_0^1 dy \int_0^{1-y} dz \left[\frac{g^{\mu\nu}}{4} J(D, 0, 2, b^2) \right. \\
&\qquad\qquad \left. + (y p_2^\mu - z p_3^\mu)(y p_2^\nu - z p_3^\nu) J(D, 0, 3, b^2) \right]
\end{aligned}
$$

$$= \frac{-i}{(4\pi)^2} \mu^{2\epsilon} \left\{ g^{\mu\nu} \frac{1}{4\hat{\epsilon}} + \int_0^1 dy \int_0^{1-y} dz \right.$$

$$\left. \times \left[\frac{1}{b^2} \left(y p_2^\mu - z p_3^\mu \right) \left(y p_2^\nu - z p_3^\nu \right) + \frac{1}{2} g^{\mu\nu} \ln \left(\frac{b^2}{\mu^2} \right) \right] \right\}. \quad (8.42)$$

For a generic N-point function one can use the Feynman parametrization (6.27) introduced previously in Chap. 6.

We will now move on to the analysis of the IR divergences of the three-point functions we have studied in this section.

8.4   IR Divergences and the Three-Point Function

Consider the following configuration for the previously introduced three point functions

$$p_2^2 = m_2^2 = m^2, \qquad p_3^2 = m_3^2 = \overline{m}^2, \qquad M^2 = 0. \quad (8.43)$$

with $m_i^2 \neq 0$ or $m_i^2 = 0$. In this case the denominator we have is:

$$\frac{1}{k^2(k^2 + 2p_2 \cdot k)(k^2 - 2p_3 \cdot k)} \xrightarrow{k^2 \ll} \frac{1}{k^2(2p_2 \cdot k)(2p_3 \cdot k)}, \quad (8.44)$$

with the two usual infrared divergences:

$$\begin{array}{lll}
1. \ k \to 0, & & \text{soft divergence,} \\
2. \ p_3 \cdot k \text{ or } p_2 \cdot k \to 0, & & \text{collinear divergence.}
\end{array} \quad (8.45)$$

Here we shall only consider the most simple case for massless on-shell particles $m_i^2 = 0$. The case with $m_i^2 \neq 0$ is left for the reader as an exercise. For these calculations we shall switch to Feynman parametrisation (6.24). Choosing $A = k^2 + 2p_2 \cdot k$, $B = k^2 - 2p_3 \cdot k$ and $C = k^2$ as in (6.34), we obtain:

$$\frac{1}{ABC} = \int_0^1 dx \int_0^1 dy \frac{2x}{\mathcal{D}^3}, \quad (8.46)$$

with the denominator $\mathcal{D}$ given by

$$\mathcal{D} = \left(k + p_2 xy - p_3 x(1 - y) \right)^2 - b^2, \quad (8.47)$$

$$b^2 = -2(p_2 \cdot p_3)x^2 y(1 - y). \quad (8.48)$$

As we have seen in Chap. 6, we can not take $D \to 4$ in $J(D, 0, 3, b^2)$ when IR divergences are present. The expression we will be needing is:

$$J(D, 0, 3, b^2) = \frac{-i}{(4\pi)^{D/2}} \frac{1}{2} \Gamma(3 - D/2)(b^2)^{D/2-3}, \tag{8.49}$$

which we have already introduced in (6.35).

The three point functions treated in this section will carry an IR sub-index. The first scalar three-point function that we have is (6.31). For completeness we shall present parts of this the calculation again:

$$
\begin{aligned}
C_{1,\text{IR}} &= 2 \int_0^1 dx \int_0^1 dy\, x\, J(D, 0, 3, b^2) \\
&= \frac{-i}{(4\pi)^{D/2}} \Gamma(3 - D/2) \int_0^1 dx \int_0^1 dy\, x (b^2)^{D/2-3} \\
&= \frac{-i}{(4\pi)^{D/2}} \Gamma(3 - D/2)(-2p_2 \cdot p_3)^{D/2-3} \int_0^1 dx\, x^{D-5} \\
&\qquad\qquad \times \int_0^1 dy\, y^{D/2-3}(1 - y)^{D/2-3} \\
&= \frac{-i}{(4\pi)^{D/2}} \frac{\Gamma(3 - D/2)}{\Gamma(D - 3)}(-2p_2 \cdot p_3)^{D/2-3} \Gamma(D/2 - 2)\Gamma(D/2 - 2).
\end{aligned}
\tag{8.50}
$$

Taking D adequately we get the following result:

$$
\begin{aligned}
C_{1,\text{IR}} &= \int \frac{d^D k}{(2\pi)^D} \frac{1}{k^2(k + p_2)^2(k - p_3)^2} \\
&= \frac{i}{(4\pi)^2} \frac{1}{2p_2 \cdot p_3} \left(\frac{-2p_2 \cdot p_3}{4\pi}\right)^{\epsilon'} \frac{\Gamma(1 - \epsilon')}{\Gamma(1 + 2\epsilon')} \Gamma^2(\epsilon')
\end{aligned}
\tag{8.51}
$$

Let's now continue and analyse the second integral $C_{2,\text{IR}}^{\mu}$:

$$
\begin{aligned}
C_{2,\text{IR}}^{\mu} &= 2 \int_0^1 dx \int_0^1 dy\, x\left(- p_2^{\mu} xy + p_3^{\mu} x(1 - y)\right) J(D, 0, 3, b^2) \\
&= \frac{-i}{(4\pi)^{D/2}} \frac{\Gamma(3 - D/2)\Gamma(D - 3)}{\Gamma(D - 2)}(-2p_2 \cdot p_3)^{D/2-3} \\
&\qquad\qquad \times \int_0^1 dy\left(- p_2^{\mu} y + p_3^{\mu}(1 - y)\right)\left(y(1 - y)\right)^{D/2-3} \\
&= \frac{-i}{(4\pi)^{D/2}} \frac{\Gamma(3 - D/2)\Gamma(D - 3)}{\Gamma(D - 2)}(-2p_2 \cdot p_3)^{D/2-3} \\
&\qquad\qquad \times (p_3^{\mu} - p_2^{\mu})\frac{\Gamma(D/2 - 1)\Gamma(D/2 - 2)}{\Gamma(D - 3)}.
\end{aligned}
\tag{8.52}
$$

We obtain the following simple result:

$$
\begin{aligned}
C^{\mu}_{2,\mathrm{IR}} &= \int \frac{d^D k}{(2\pi)^D} \frac{k^{\mu}}{k^2(k+p_2)^2(k-p_3)^2} \\
&= (p_3^{\mu} - p_2^{\mu}) \frac{i}{(4\pi)^2} \frac{1}{2p_2 \cdot p_3} \\
&\quad \times \left(\frac{-2p_2 \cdot p_3}{4\pi} \right)^{\epsilon'} \frac{\Gamma(1-\epsilon')\Gamma(1+\epsilon')}{\Gamma(2+2\epsilon')} \Gamma(\epsilon')
\end{aligned}
\tag{8.53}
$$

Continuing with the third integral we get:

$$
\begin{aligned}
C_{3,\mathrm{IR}} &= \int_0^1 dx \int_0^1 dy \int \frac{d^D k}{(2\pi)^D} \frac{2xk^2}{[(k+p_2 xy - p_3 x(1-y))^2 - b^2]^3} \\
&= \int_0^1 dx \int_0^1 dy \left(\frac{x}{2} D\, J(D,0,2,b^2) - 4(p_2 \cdot p_3)x^3 y(1-y) J(D,0,3,b^2) \right) \\
&\equiv \mathcal{B}\big(J(D,0,2,b^2)\big) + \mathcal{C}\big(J(D,0,3,b^2)\big).
\end{aligned}
\tag{8.54}
$$

The $\mathcal{B}$ function is, of course, UV divergent:

$$
\begin{aligned}
\mathcal{B} &= \int_0^1 dx \int_0^1 dy \frac{x}{2} D\, J(D,0,2,b^2) \\
&= \frac{-i}{(4\pi)^2} \mu^{2\epsilon} \left[\frac{1}{\hat{\epsilon}} + \frac{1}{2} + 2 \int_0^1 dx \int_0^1 dy\, x \ln\left(\frac{b^2}{\mu^2} \right) \right] \\
&= \frac{-i}{(4\pi)^2} \mu^{2\epsilon} \left[\frac{1}{\hat{\epsilon}} + \frac{1}{2} + \ln\left(\frac{-2p_2 \cdot p_3}{\mu^2} \right) + 2 \int_0^1 dx \int_0^1 dy\, x \ln[x^2 y(1-y)] \right] \\
&= \frac{-i}{(4\pi)^2} \mu^{2\epsilon} \left[\frac{1}{\hat{\epsilon}} - \frac{5}{2} + \ln\left(\frac{2p_2 \cdot p_3}{\mu^2} \right) \pm i\pi \right],
\end{aligned}
\tag{8.55}
$$

where we have taken

$$
\ln(-1) = \ln e^{\pm i\pi} = \pm i\pi.
\tag{8.56}
$$

The second integral, as we shall see is in fact, finite:

$$
\begin{aligned}
\mathcal{C} &= -4(p_2 \cdot p_3) \int_0^1 dx \int_0^1 dy\, x^3 y(1-y) J(D,0,3,b^2) \\
&= 2(p_2 \cdot p_3) \frac{i}{(4\pi)^{D/2}} \Gamma(3 - D/2) \int_0^1 dx \int_0^1 dy\, x^3 y(1-y)(b^2)^{D/2-3}.
\end{aligned}
\tag{8.57}
$$

We have no poles at $D \to 4$ thus, we obtain the simple result

$$C = \frac{-i}{(4\pi)^2} \frac{1}{2} + O(\epsilon'). \tag{8.58}$$

Therefore, our scalar integral is finally given by

$$\boxed{\begin{aligned}
C_{3,\mathrm{IR}} &= \int \frac{d^D k}{(2\pi)^D} \frac{k^2}{k^2 (k + p_2)^2 (k - p_3)^2} \\
&= \frac{-i}{(4\pi)^2} \mu^{2\epsilon} \left[\frac{1}{\hat{\epsilon}} - 2 + \ln\left(\frac{2 p_2 \cdot p_3}{\mu^2}\right) \pm i\pi \right]
\end{aligned}} \tag{8.59}$$

The last integral we are going to analyse is $C_{4,\mathrm{IR}}^{\mu\nu}$:

$$\begin{aligned}
C_{4,\mathrm{IR}}^{\mu\nu} &= \int_0^1 dx \int_0^1 dy \int \frac{d^D k}{(2\pi)^D} \frac{2x}{[k^2 - b^2]^3} \\
&\quad \times \left[k^\mu - p_2^\mu x y + p_3^\mu x(1 - y) \right] \left[k^\nu - p_2^\nu x y + p_3^\nu x(1 - y) \right] \\
&= \int_0^1 dx \int_0^1 dy \left[g^{\mu\nu} \frac{x}{2} J(D, 0, 2, b^2) + 2x^3 \left(p_2^\mu y - p_3^\mu (1 - y) \right) \right. \\
&\quad \left. \times \left(p_2^\nu y - p_3^\nu (1 - y) \right) J(D, 0, 3, b^2) \right] \\
&\equiv \mathcal{E}^{\mu\nu} \left(J(D, 0, 2, b^2) \right) + \mathcal{F}^{\mu\nu} \left(J(D, 0, 3, b^2) \right).
\end{aligned} \tag{8.60}$$

The UV-divergent tensor function $\mathcal{E}^{\mu\nu}$ gives:

$$\begin{aligned}
\mathcal{E}^{\mu\nu} &= \int_0^1 dx \int_0^1 dy\, g^{\mu\nu} \frac{x}{2} J(D, 0, 2, b^2) \\
&= g^{\mu\nu} \frac{-i}{(4\pi)^2} \mu^{2\epsilon} \int_0^1 dx \int_0^1 dy \frac{x}{2} \left[\frac{1}{\hat{\epsilon}} + \ln\left(\frac{b^2}{\mu^2}\right) \right] \\
&= g^{\mu\nu} \frac{-i}{(4\pi)^2} \mu^{2\epsilon} \frac{1}{4} \left[\frac{1}{\hat{\epsilon}} + \ln\left(\frac{2 p_2 \cdot p_3}{\mu^2}\right) - 3 \pm i\pi \right].
\end{aligned} \tag{8.61}$$

The remaining structure is IR divergent:

$$\begin{aligned}
\mathcal{F}^{\mu\nu} &= \int_0^1 dx \int_0^1 dy\, 2x^3 \left(p_2^\mu y - p_3^\mu (1 - y) \right) \\
&\quad \times \left(p_2^\nu y - p_3^\nu (1 - y) \right) J(D, 0, 3, b^2) \\
&= \frac{-i}{(4\pi)^{D/2}} \frac{\Gamma(3 - D/2)\Gamma(D - 2)}{\Gamma(D - 1)} \int_0^1 dy \left(p_2^\mu y - p_3^\mu (1 - y) \right) \\
&\quad \times \left(p_2^\nu y - p_3^\nu (1 - y) \right) \left(-2 p_2 \cdot p_3 y(1 - y) \right)^{D/2 - 3}
\end{aligned}$$

$$= \frac{-i}{(4\pi)^{D/2}} \frac{\Gamma(3 - D/2)\Gamma(D - 2)}{\Gamma(D - 1)} (-2p_2 \cdot p_3)^{D/2-3}$$

$$\times \left[(p_2^\mu p_2^\nu + p_3^\mu p_3^\nu) \frac{\Gamma(D/2)\Gamma(D/2 - 2)}{\Gamma(D - 2)} \right.$$

$$\left. - (p_2^\mu p_3^\nu + p_2^\nu p_3^\mu) \frac{\Gamma(D/2 - 1)\Gamma(D/2 - 1)}{\Gamma(D - 2)} \right]. \tag{8.62}$$

Therefore, the final expression for $C_{4,\mathrm{IR}}^{\mu\nu}$ that we find is:

$$C_{4,\mathrm{IR}}^{\mu\nu} = \int \frac{d^D k}{(2\pi)^D} \frac{k^\mu k^\nu}{k^2 (k + p_2)^2 (k - p_3)^2}$$

$$= g^{\mu\nu} \frac{-i}{(4\pi)^2} \mu^{2\epsilon} \frac{1}{4} \left[\frac{1}{\epsilon} + \ln\left(\frac{2p_2 \cdot p_3}{\mu^2} \right) - 3 \pm i\pi \right]$$

$$+ \frac{i}{(4\pi)^2} \frac{1}{2p_2 \cdot p_3} \left(\frac{-2p_2 \cdot p_3}{4\pi} \right)^{\epsilon'} \frac{\Gamma(1 - \epsilon')}{\Gamma(3 + 2\epsilon')}$$

$$\times \left[(p_2^\mu p_2^\nu + p_3^\mu p_3^\nu) \Gamma(2 + \epsilon') \Gamma(\epsilon') \right.$$

$$\left. - (p_2^\mu p_3^\nu + p_2^\nu p_3^\mu) \Gamma^2(1 + \epsilon') \right] \tag{8.63}$$

In the following we shall introduce the two and three-body phase space expressions in $D = 4 + 2\epsilon'$ dimensions, needed for the calculation of IR-divergent cross sections. Afterwards, we will calculate an IR divergent process for which we will make use of all the IR-divergent integrals introduced in this section and also the IR-divergent two-point function presented previously.

8.5 Two and Three-Body Phase Space in D Dimensions

For $D = 4 + 2\epsilon'$ with $\epsilon' > 0$ we have the following two-body phase space:

$$\int dQ_2 = \frac{1}{4\pi} \frac{\hat{p}}{\sqrt{s}} \left(\frac{\hat{p}^2}{\pi} \right)^{\epsilon'} \frac{1}{\Gamma(1 + \epsilon')} \int_0^1 dv [v(1 - v)]^{\epsilon'} \tag{8.64}$$

where $\hat{p} \equiv |\mathbf{p}| = \gamma m |\mathbf{v}|$ is the center of mass momentum (of the final state particles) given by

$$\hat{p} = \frac{1}{2\sqrt{s}} \lambda^{1/2}(s, m_a^2, m_b^2), \tag{8.65}$$

and where $\lambda(x, y, z)$ is the usual Kallen function $\lambda(x, y, x) \equiv x^2 + y^2 + z^2 - 2xy - 2yz - 2xz$. The variable v is defined as

$$v = \frac{1 + \cos\theta}{2},$$ (8.66)

with θ the CM scattering angle. If we are dealing with final state identical particles, we need to also multiply by the $1/2$ symmetry factor. If the squared transition matrix does not depend on the angle θ, we can integrate over v and obtain the simple result

$$\boxed{\int dQ_2 = \frac{1}{4\pi} \frac{\hat{p}}{\sqrt{s}} \left(\frac{\hat{p}^2}{\pi}\right)^{\epsilon'} \frac{\Gamma(1 + \epsilon')}{\Gamma(2 + 2\epsilon')}.}$$ (8.67)

The three-body phase space is given by:

$$\boxed{\int dQ_3 = \frac{Q^2}{2(4\pi)^3}\left(\frac{4\pi}{Q^2}\right)^{-2\epsilon'} \frac{1}{\Gamma(2 + 2\epsilon')} \int_0^1 dx_1 \int_{1-x_1}^1 dx_2 \\ \times \left[(1 - x_1)(1 - x_2)(1 - x_3)\right]^{\epsilon'}}$$ (8.68)

where $Q^\mu = p_1^\mu + p_2^\mu + p_3^\mu$ (with p_i^μ the final state particle momenta) and where we have introduced the x_i kinematical variables:

$$x_1 \equiv \frac{2(p_1 \cdot Q)}{Q^2}, \qquad x_2 \equiv \frac{2(p_2 \cdot Q)}{Q^2}, \qquad x_3 \equiv \frac{2(p_3 \cdot Q)}{Q^2}.$$ (8.69)

It is easy to check that $x_1 + x_2 + x_3 = 2$. If we have n final state identical particles ($n = 2, 3$) we must also multiply by the $\frac{1}{n!}$ symmetry factor.

When averaging over the spins of the initial state particles, it is conventional to consider for massless on-shell photons or gluons $D - 2 = 2(1 + \epsilon')$ polarizations. Fermions are considered to have two spin-polarization states as usual.

8.6   Cancellation of IR Divergences

Consider the following collision $f\bar{f} \rightarrow \gamma^* \rightarrow e^+e^-\gamma$, with $f\bar{f}$ any fermion-antifermion pair. The corresponding cross-section for this process can be written as

$$\sigma \sim F^{\mu\nu} \int dQ_3 \, L_{\mu\nu},$$ (8.70)

with $F^{\mu\nu}$ the initial state fermionic tensor and with $L_{\mu\nu}$ the final state leptonic tensor. Due to QED gauge invariance we can write the previous expression (8.70) in the form

$$\sigma \sim F^{\alpha\beta} \, g_{\alpha\beta} \int dQ_3 \, L_{\mu\nu} \, g^{\mu\nu}, \qquad (8.71)$$

therefore, the calculation of this cross section can be broken into two disjoint pieces. As we are only interested in the final state (it is where the IR divergences will appear) we will just focus on the sub-process $\gamma^* \to e^+ e^- \gamma$. Thus, for practical purposes we can consider that the virtual photon γ^* is described by some polarization vectors ϵ_s^{μ} and, we can perform the usual substitution $\sum_s \epsilon_s^{\mu} \epsilon_s^{\nu*} \to -g^{\mu\nu}$ when calculating the squared transition matrix.[2] For this analysis we shall consider massless electrons. At leading order there are two diagrams that contribute to our sub-process:

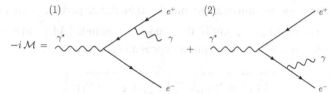

where the squared transition matrix $|\mathcal{M}|^2 = |\mathcal{M}_1 + \mathcal{M}_2|^2$ is of $\mathcal{O}(e^4)$. When calculating the cross section corresponding to this process one finds that it has IR divergences. This is mainly due to the fact that, we can have soft or collinear emitted photons that are experimentally indistinguishable from just a electron. Schematically:

soft   collinear

$\longrightarrow$ or $\longrightarrow$ $\sim$ $\longrightarrow$

Roughly speaking, in the soft or collinear limit the three-body final state *turns into* a two-body final state. Thus, in order to be able to remove the IR divergences we must also analyse the $2 \to 2$ scattering process given by $f\bar{f} \to \gamma^* \to e^+ e^-$ (with $f\bar{f}$ the same fermion-antifermion pair as in the previous case). The cross section for this process can be separated exactly as in (8.71) (obviously with dQ_2 instead of dQ_3) therefore we can, again, focus only on the sub-process $\gamma^* \to e^+ e^-$. The transition amplitude for this sub-process including one-loop corrections (see Chap. 7) it is given by:

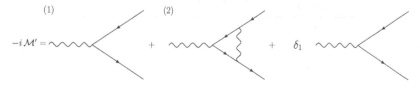

---

[2] As the photon is not a *real* (on-shell) particle we shall not average over its polarizations when calculating the cross section.

The matrix element can, thus, be written as $\mathcal{M}' = \mathcal{M}'_1 + \mathcal{M}'_2 + \delta_1 \mathcal{M}'_1$ with the amplitude $\mathcal{M}'_1$ of $\mathcal{O}(e)$ and with the amplitudes $\mathcal{M}'_2$ and $\delta_1 \mathcal{M}'_1$ of $\mathcal{O}(e^3)$. As we already know δ_1 is an IR divergent quantity and, as we can suspect, it will also be the case for the amplitude $\mathcal{M}'_2$. It turns out that summing to the cross section of the first process $\sigma(\gamma^* \to e^+ e^- \gamma)$, the part of $\sigma(\gamma^* \to e^+ e^-)$ that contributes to the same order in perturbation theory, that is $\mathcal{O}(e^4)$, the IR poles cancel. We can therefore define a proper infrared safe (IRS) observable given by[3]

$$
\boxed{
\begin{aligned}
\sigma^{\mathrm{IRS}}(\gamma^* \to e^+ e^- \gamma) &= \sigma(\gamma^* \to e^+ e^- \gamma) + \bar{\sigma}(\gamma^* \to e^+ e^-) \\
&= \frac{1}{2s} \left(\int d\mathcal{Q}_3 \sum_{r_i} |\mathcal{M}|^2 + \int d\mathcal{Q}_2 \sum_{r_i} |\bar{\mathcal{M}}'|^2 \right)
\end{aligned}
}
\tag{8.72}
$$

where $\sum_{r_i}$ stands for the sum over the initial and final state particle polarizations and where $|\bar{\mathcal{M}}'|^2$ is defined as the part of the squared amplitude $|\mathcal{M}'|^2$, that contributes to the cross section at $\mathcal{O}(e^4)$. Its explicit expression is given by

$$
\boxed{
|\bar{\mathcal{M}}'|^2 \equiv 2\mathrm{Re}[\mathcal{M}'^{\dagger}_1 \mathcal{M}'_2] + 2\delta_1 |\mathcal{M}'_1|^2
}
\tag{8.73}
$$

Let's thus, start by calculating the first process:

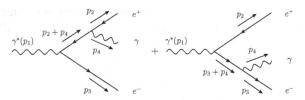

The squared matrix element is given by $|\mathcal{M}|^2 = |\mathcal{M}_1 + \mathcal{M}_2|^2 = |\mathcal{M}_1|^2 + |\mathcal{M}_2|^2 + 2\mathrm{Re}[\mathcal{M}^{\dagger}_1 \mathcal{M}_2]$. In D dimensions we obtain the following results:

$$
\boxed{
\begin{aligned}
\sum_{r_i} |\mathcal{M}_1|^2 &= \frac{e^4}{4(p_2 \cdot p_4)^2} \mathrm{Tr}\{\gamma^{\nu}(\not{p}_2 + \not{p}_4)\gamma^{\mu} \not{p}_3 \gamma_{\mu}(\not{p}_2 + \not{p}_4)\gamma_{\nu} \not{p}_2\} \\
&= \frac{2e^4}{(p_2 \cdot p_4)}(D - 2)^2 (p_3 \cdot p_4)
\end{aligned}
}
\tag{8.74}
$$

---

[3] Read Chap. 3 for Relativistic kinematics, phase space and cross section formulae, and Chap. 5 for more details on calculations of QED processes and the corresponding Feynman rules.

$$\sum_{r_i} |\mathcal{M}_2|^2 = \frac{e^4}{4(p_3 \cdot p_4)^2} \text{Tr}\{\gamma^\mu (\not{p}_3 + \not{p}_4)\gamma^\nu \not{p}_3 \gamma_\nu (\not{p}_3 + \not{p}_4)\gamma_\mu \not{p}_2\}$$

$$= \frac{2e^4}{(p_3 \cdot p_4)}(D-2)^2 (p_2 \cdot p_4) \tag{8.75}$$

The crossed term reads:

$$\sum_{r_i} 2\text{Re}[\mathcal{M}_1^\dagger \mathcal{M}_2] = -\frac{e^4}{4(p_2 \cdot p_4)(p_3 \cdot p_4)}$$

$$\times \text{Tr}\{\gamma^\nu (\not{p}_2 + \not{p}_4)\gamma^\mu \not{p}_3 \gamma_\nu (\not{p}_3 + \not{p}_4)\gamma_\mu \not{p}_2\}$$

$$= \frac{4e^4}{(p_3 \cdot p_4)(p_2 \cdot p_4)}(D-2)$$

$$\times \Big[(D-4)(p_2 \cdot p_4)(p_3 \cdot p_4)$$

$$+ 2(p_2 \cdot p_3)^2 + 2(p_2 \cdot p_4)(p_2 \cdot p_3)$$

$$+ 2(p_3 \cdot p_4)(p_2 \cdot p_3) \Big] \tag{8.76}$$

The minus sign (from the RHS of the first line of the previous expression) is due to the fermionic propagator of the first diagram (which goes in the opposite direction of the momentum, and therefore it has the form $-i(\not{p}_2 + \not{p}_4)/(2p_2 \cdot p_4)$ and not $i(\not{p}_2 + \not{p}_4)/(2p_2 \cdot p_4)$). Introducing the x_i kinematical variables from the previous section

$$x_1 \equiv \frac{2(p_2 \cdot Q)}{Q^2}, \qquad x_2 \equiv \frac{2(p_3 \cdot Q)}{Q^2}, \qquad x_3 \equiv \frac{2(p_4 \cdot Q)}{Q^2}, \tag{8.77}$$

with $Q \equiv p_1 = p_2 + p_3 + p_4$ (and where, remember, $x_1 + x_2 + x_3 = 2$) we obtain the following results:

$$(p_2 \cdot p_4) = \frac{1}{2}x_1 Q^2 - (p_2 \cdot p_3), \tag{8.78}$$

$$(p_3 \cdot p_4) = \frac{1}{2}x_2 Q^2 - (p_2 \cdot p_3), \tag{8.79}$$

$$(p_2 \cdot p_3) = \frac{1}{2}Q^2(x_1 + x_2 - 1). \tag{8.80}$$

Performing these substitutions, taking $D = 4 + 2\epsilon'$ (with $\epsilon' > 0$) and keeping the terms up to $O(\epsilon'^2)$ we obtain the following simple formula for the squared transition matrix

$$\sum_{r_i} |\mathcal{M}|^2 = 8e^4(\epsilon' + 1)\frac{x_1^2 + x_2^2 + \epsilon'(x_1 + x_2 - 2)^2}{(1 - x_1)(1 - x_2)}. \tag{8.81}$$

Introducing the three-body phase space in D dimensions (8.68) we have:

$$
\int dQ_3 \sum_{r_i} |\mathcal{M}|^2 = \frac{4e^4 Q^2 (\epsilon'+1)}{(4\pi)^3} \left(\frac{4\pi}{Q^2}\right)^{-2\epsilon'} \frac{1}{\Gamma(2+2\epsilon')} \int_0^1 dx_1 \int_{1-x_1}^1 dx_2
$$

$$
\times \frac{1}{[(1-x_1)(1-x_2)(x_1+x_2-1)]^{-\epsilon'}}
$$

$$
\times \frac{x_1^2 + x_2^2 + \epsilon'(x_1+x_2-2)^2}{(1-x_1)(1-x_2)}
$$

$$
\equiv \frac{4e^4 Q^2 (\epsilon'+1)}{(4\pi)^3} \left(\frac{4\pi}{Q^2}\right)^{-2\epsilon'} \frac{1}{\Gamma(2+2\epsilon')} K(\epsilon') \qquad (8.82)
$$

The remaining task is to calculate the integral $K(\epsilon')$. Performing a change of variables $x_1 = x$ and $x_2 = 1 - vx$ we obtain:

$$
K(\epsilon') = \int_0^1 dx \int_0^1 dv \frac{x}{[v(1-v)x^2(1-x)]^{-\epsilon'}}
$$

$$
\times \frac{v^2 x^2 - 2vx + x^2 + 1 + \epsilon'(x(1-v)-1)^2}{vx(1-x)}
$$

$$
= \int_0^1 dx \int_0^1 dv \frac{x}{[v(1-v)x^2(1-x)]^{-\epsilon'}}
$$

$$
\times \left[(1+\epsilon')\left(\frac{1-x}{vx} + \frac{vx}{1-x}\right) + \frac{2(1-v)}{v(1-x)} + 2\epsilon' \right]
$$

$$
= \int_0^1 dx \int_0^1 dv \left[v^{\epsilon'}(1-v)^{\epsilon'} x^{2\epsilon'+1} (1-x)^{\epsilon'} \right] \times
$$

$$
\times \left[\left(x^{-1}(1-x)v^{-1} + vx(1-x)^{-1}\right)(1+\epsilon') \right.
$$

$$
\left. + 2v^{-1}(1-v)(1-x)^{-1} + 2\epsilon' \right]
$$

$$
= \frac{2(1+\epsilon')\Gamma(\epsilon')\Gamma(\epsilon'+1)\Gamma(\epsilon'+2)}{\Gamma(3\epsilon'+3)} + \frac{2\Gamma^2(\epsilon')\Gamma(\epsilon'+2)}{\Gamma(3\epsilon'+2)} + \frac{2\epsilon'\Gamma^3(1+\epsilon')}{\Gamma(3\epsilon'+3)}
$$

$$
= \frac{2}{\epsilon'^2} - \frac{3}{\epsilon'} + \frac{19}{2} - \pi^2. \qquad (8.83)
$$

Thus, we have found the following result for the cross section of first process:

$$
2s\,\sigma(\gamma^* \to e^+ e^- \gamma) = \int dQ_3 \sum_{r_i} |\mathcal{M}|^2
$$

$$
= \frac{\alpha^2 Q^2}{\pi} \left(\frac{4\pi}{Q^2}\right)^{-2\epsilon'} \frac{(\epsilon'+1)}{\Gamma(2+2\epsilon')}
$$

$$
\times \left(\frac{2}{\epsilon'^2} - \frac{3}{\epsilon'} + \frac{19}{2} - \pi^2\right) \qquad (8.84)
$$

We now move on, and calculate the remaining cross section $\bar{\sigma}(\gamma^* \to e^+ e^-)$. The first amplitude that contributes to this process is given by $\mathcal{M}'_1$:

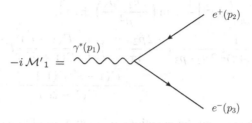

$$\mathcal{M}'_1 = -e\epsilon_\mu^{r_1}(p_1)\bar{u}_{r_3}(p_3)\gamma^\mu v_{r_2}(p_2). \tag{8.85}$$

where we have taken the charge of the electron $Q = -1$ (do not confuse it with the four-momentum Q^μ). The second amplitude of the process, $\mathcal{M}'_2$ is given by:

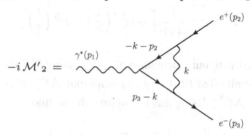

$$\mathcal{M}'_2 = i\, e^3 \epsilon_\mu^{r_1}(p_1)\bar{u}_{r_3}(p_3)\Gamma^\mu v_{r_2}(p_2), \tag{8.86}$$

where Γ^μ reads

$$\Gamma^\mu = \int \frac{d^D k}{(2\pi)^D} \frac{\gamma^\rho(\slashed{k} - \slashed{p}_3)\gamma^\mu(\slashed{k} + \slashed{p}_2)\gamma_\rho}{k^2(k + p_2)^2(k - p_3)^2}. \tag{8.87}$$

Working in D dimensions, and using the Dirac equations of motion for massless fermions $\slashed{p}_2 v_{r_2}(p_2) = 0$ and $\bar{u}_{r_3}(p_3)\slashed{p}_3 = 0$ in order to eliminate some of the terms, we find

$$
\begin{aligned}
\Gamma^\mu &= \int \frac{d^D k}{(2\pi)^D} \frac{(4 - 2D)k^\mu \slashed{k} + (D - 2)\gamma^\mu k^2 + 2\slashed{k}\gamma^\mu \slashed{p}_3 - 2\slashed{p}_2\gamma^\mu \slashed{k} + 2\slashed{p}_2\gamma^\mu \slashed{p}_3}{k^2(k + p_2)^2(k - p_3)^2} \\
&= (4 - 2D)\gamma_\nu C_{4,\mathrm{IR}}^{\mu\nu} + (D - 2)\gamma^\mu C_{3,\mathrm{IR}} \\
&\qquad + 2(\gamma_\nu \gamma^\mu \slashed{p}_3 - \slashed{p}_2\gamma^\mu \gamma_\nu)C_{2,\mathrm{IR}}^\nu + 2\slashed{p}_2\gamma^\mu \slashed{p}_3 C_{1,\mathrm{IR}} \\
&= A\gamma^\mu + B\slashed{p}_2\gamma^\mu \slashed{p}_3.
\end{aligned} \tag{8.88}
$$

The coefficients A and B are given by

$$A = \frac{-i}{(4\pi)^2} \mu^{2\epsilon} \left[\frac{1}{\hat{\epsilon}} + \ln\left(\frac{2p_2 \cdot p_3}{\mu^2} \right) \pm i\pi \right],$$

$$B = \frac{i}{(4\pi)^2} \frac{1}{2p_2 \cdot p_3} \left(\frac{-2p_2 \cdot p_3}{4\pi} \right)^{\epsilon} \left[2 \frac{\Gamma(1-\epsilon')}{\Gamma(1+2\epsilon')} \Gamma^2(\epsilon') \right.$$
$$\left. - 4\frac{\Gamma(1-\epsilon')\Gamma(1+\epsilon')}{\Gamma(2+2\epsilon')} \Gamma(\epsilon') \right], \qquad (8.89)$$

where again, we have used the equations of motion to eliminate terms. The last amplitude is simply $-i\delta_1 \mathcal{M}'_1$ with δ_1 given by (8.29). We obtain

$$\delta_1 \mathcal{M}'_1 = i e^3 \epsilon_\mu^{r_1}(p_1) \bar{u}_{r_3}(p_3) \tilde{\Gamma}^\mu v_{r_2}(p_2), \qquad (8.90)$$

where we have defined $\tilde{\Gamma}^\mu$ as

$$\tilde{\Gamma}^\mu = \gamma^\mu \frac{i}{(4\pi)^2} \left[-\mu^{2\epsilon'} \left(\frac{1}{\hat{\epsilon}'} \right) + \mu^{2\epsilon} \left(\frac{1}{\hat{\epsilon}} \right) \right]. \qquad (8.91)$$

We can now proceed with our calculation of $|\bar{\mathcal{M}}'|^2$ using (8.73) or equivalently, we can define the UV-finite (but IR-divergent) amplitude $\mathcal{M}'^R_2 = \mathcal{M}'_2 + \delta_1 \mathcal{M}'_1$ thus, $|\bar{\mathcal{M}}'|^2 = 2\text{Re}\,[\mathcal{M}'^\dagger_1 \mathcal{M}'^R_2]$. Using this last approach, we find

$$\mathcal{M}'^R_2 = i\,e^3 \epsilon_\mu^{r_1}(p_1) \bar{u}_{r_3}(p_3) \Gamma^\mu_R v_{r_2}(p_2), \qquad (8.92)$$

with $\Gamma^\mu_R = \Gamma^\mu + \tilde{\Gamma}^\mu = A_R \gamma^\mu + B \not{p}_2 \gamma^\mu \not{p}_3$, and where

$$A_R = \frac{-i}{(4\pi)^2} \left[\mu^{2\epsilon'} \frac{1}{\hat{\epsilon}'} + \ln\left(\frac{2p_2 \cdot p_3}{\mu^2} \right) \pm i\pi \right]$$
$$= \frac{-i}{(4\pi)^2} \left[\frac{1}{\epsilon'} + \gamma_E - \ln(4\pi) + \ln(2p_2 \cdot p_3) \pm i\pi \right] + O(\epsilon). \qquad (8.93)$$

Using the expansion

$$\ln\left(\frac{2p_2 \cdot p_3}{4\pi} \right) = \frac{1}{\epsilon'} \left[\left(\frac{2p_2 \cdot p_3}{4\pi} \right)^{\epsilon'} - 1 \right] + O(\epsilon'), \qquad (8.94)$$

the final expression of A_R reads

$$A_R = \frac{-i}{(4\pi)^2} \left[\frac{1}{\epsilon'} \left(\frac{2p_2 \cdot p_3}{4\pi} \right)^{\epsilon'} + \gamma_E \pm i\pi \right]. \qquad (8.95)$$

The squared amplitude is finally given by:

$$
\sum_{r_i} |\bar{\mathcal{M}}'|^2 = 2\,\mathrm{Re}\Big[i\,e^4\,\mathrm{Tr}\{\gamma_\mu \not{p}_3\,(\gamma^\mu A_R + B\not{p}_2\gamma^\mu \not{p}_3)\,\not{p}_2\}\Big]
$$

$$
= 8\,\alpha^2\,Q^2\left(\frac{Q^2}{4\pi}\right)^{\epsilon'}(1+\epsilon')
$$

$$
\times \left[\mathrm{Re}\Big((-1)^{\epsilon'}\Big)\Big(-2\frac{\Gamma(1-\epsilon')}{\Gamma(1+2\epsilon')}\Gamma^2(\epsilon')\right.
$$

$$
\left. +4\frac{\Gamma(1-\epsilon')\Gamma(1+\epsilon')}{\Gamma(2+2\epsilon')}\Gamma(\epsilon')\Big) - \frac{1}{\epsilon'} - \gamma_E\right] \qquad (8.96)
$$

where $Q^2 = p_1^2 = 2p_2 \cdot p_3$. As the squared transition matrix does not depend on the angle θ we can directly introduce the expression (8.67) for the phase space. Expanding in ϵ' we obtain the needed cross section:

$$
2s\,\bar{\sigma}(\gamma^* \to e^+e^-) = \int dQ_2 \sum_{r_i} |\bar{\mathcal{M}}'|^2
$$

$$
= \frac{\alpha^2 Q^2}{\pi}\left(\frac{4\pi}{Q^2}\right)^{-2\epsilon'}\frac{(1+\epsilon')}{\Gamma(2+2\epsilon')}\left(-\frac{2}{\epsilon'^2} + \frac{3}{\epsilon'} - 8 + \pi^2\right) \qquad (8.97)
$$

Thus, as we mentioned, the poles from the two processes cancel, and the sum is finite

$$
\sigma^{\mathrm{IRS}}(\gamma^* \to e^+e^-\gamma) = \frac{3\alpha^2}{4\pi}. \qquad (8.98)
$$

When studying examples of IR divergences cancellation, the typical process that one finds is $e^+e^- \to$ hadrons. At leading order the cross section for this process is simply given by $\sigma(e^+e^- \to q\bar{q}) \equiv \sigma_0$ where $q\bar{q}$ stands for a quark-antiquark pair. At next-to-leading order one simply adds a gluon attached to the final state quarks $\sigma(e^+e^- \to q\bar{q}g) \equiv \sigma_1$. This last process is of course IR divergent (due to radiation of soft or collinear gluons). By calculating the QCD vertex correction for the first process ($e^+e^- \to q\bar{q}$), with a virtual gluon attached to the final state quarks, again one finds IR divergences. As seen in the previous example, the IR divergences from the vertex correction of σ_0 and the ones from σ_1 cancel and one finds

$$
\sigma(e^+e^- \to \text{hadrons}) = \sigma_0\left(1 + \frac{3\alpha_s^2}{4\pi}\,C_F + \dots\right). \qquad (8.99)
$$

with α_s the strong QCD coupling and $C_F = 4/3$ the $SU(3)$ invariant. The calculation for this process is exactly the same as the one we just did, except we have made a

Fig. 8.4 Two-loop diagram
that contributes to the
anomalous magnetic
moment of the muon

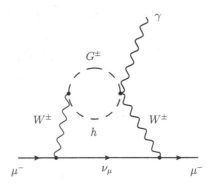

few simplifications. We have considered photons instead of gluons in the final state
(thus, we got rid of the colour factors), we have ignored the initial state collision,
and we have also ignored the tree-level process (given by σ_0).

8.7 Introduction to Two-Loops

Two-loop calculations are a highly advanced topic and one small subsection is obviously not able to cover it. However we just want to provide the reader with one last
computational tool. As we have mentioned at the beginning of this chapter, we shall
explain with a very simple two-loop example how the generic Feynman parametrization (6.27) can be used for non-integer powers of propagators.

Imagine we want to calculate the contribution from Fig. 8.4 to the anomalous
magnetic moment of the muon (or electron), where $G^\pm$ is the charged Goldstone
boson, $W^\pm$ the charged gauge boson, and h the Higgs boson of the Standard Model.[4]
This is a quite simple topology and can be broken into two parts. We will first calculate
the $\gamma W W$ effective vertex (which will, fortunately, turn out to be finite) and we will
use the expression of this vertex to calculate the second loop. Diagrammatically
this is shown in Fig. 8.5, where $Q = -1$ is the electric charge of the muon. The
amplitude of this process has the form $\mathcal{M} \sim \epsilon_\mu^s \, \Gamma^\mu$ (with ϵ_μ^s the polarization vector
of the photon) thus all terms proportional to q^μ will not contribute. The structure of
the Γ^μ vertex can therefore be written as

$$\bar{u}_r(p+q) \, \Gamma^\mu \, u_{r'}(p) = \bar{u}_r(p+q) \left[\gamma^\mu F_1(q^2) + \frac{i}{2m}\sigma^{\mu\nu}q_\nu F_2(q^2) \right.$$
$$\left. + \gamma_5 \frac{i}{2m}\sigma^{\mu\nu}q_\nu F_3(q^2) + \dots \right] u_{r'}(p). \quad (8.100)$$

---

[4]For a complete set of Feynman rules for the Standard Model read J. C. Romao and J. P. Silva, *A
resource for signs and Feynman diagrams of the Standard Model*, Int. J. Mod. Phys. A **27** (2012)
1230025, http://arxiv.org/pdf/1209.6213.pdf.

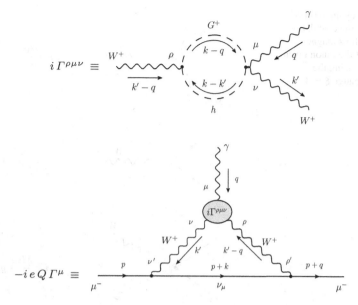

Fig. 8.5 One-loop γWW effective vertex and its insertion into the second loop

with $F_i(q^2)$ scalar form factors and where m is the mass of the muon. The anomalous magnetic moment of the muon, denoted as Δa_μ, is given by the $F_2(q^2)$ form factor for an on-shell photon i.e., $\Delta a_\mu = F_2(q^2 = 0)$. Using the standard Feynman rules for the electro-weak SM Lagrangian, working in the linear R_ξ gauge with $\xi_W = 1$ (thus $M_{G^\pm} = M_W$), one finds (the needed Feynman rules for this calculation are given in Fig. 8.6)

$$i\,\Gamma^{\rho\mu\nu} = -i\,\frac{e}{(4\pi)^2}\,\frac{M_W^2}{v^2}\,g^{\mu\nu}(k'^\rho - q^\rho)\,\mu^{2\epsilon}\int_0^1 dx\,(1-2x)\left(\frac{1}{\hat{\epsilon}} + \ln\frac{a^2}{\mu^2}\right),$$
(8.101)

with $a^2 = -x(1-x)(k'^2 - M_x^2 - 2k'\cdot q)$ and with $M_x^2 = M_h^2/(1-x) + M_W^2/x$. It is obvious from (8.101) that, when integrating over x, the divergence (and hence the μ-dependence) vanishes. Thus we are left with a finite effective vertex given by

$$i\,\Gamma^{\rho\mu\nu} = -i\,\frac{e}{(4\pi)^2}\,\frac{M_W^2}{v^2}\,g^{\mu\nu}(k'^\rho - q^\rho)\int_0^1 dx\,(1-2x)\,\ln a^2.$$
(8.102)

Using the expansion (8.94), but this time for an arbitrary small parameter δ ($\delta \ll 1$)

$$\ln a^2 = \frac{1}{\delta}\left[(a^2)^\delta - 1\right] + O(\delta),$$
(8.103)

Fig. 8.6 Feynman rules needed for the calculation of the anomalous magnetic moment of the muon given in Fig. 8.4, using the Feynman gauge $\xi = 1$

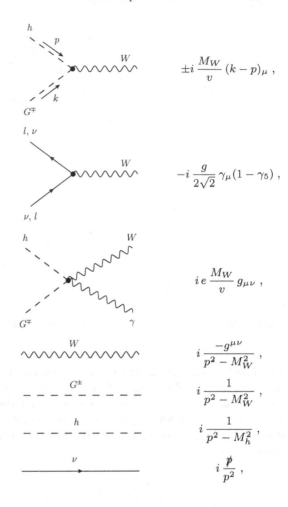

$$\pm i \, \frac{M_W}{v} \, (k - p)_\mu \, ,$$

$$-i \, \frac{g}{2\sqrt{2}} \, \gamma_\mu (1 - \gamma_5) \, ,$$

$$i \, e \, \frac{M_W}{v} \, g_{\mu\nu} \, ,$$

$$i \, \frac{-g^{\mu\nu}}{p^2 - M_W^2} \, ,$$

$$i \, \frac{1}{p^2 - M_W^2} \, ,$$

$$i \, \frac{1}{p^2 - M_h^2} \, ,$$

$$i \, \frac{\not{p}}{p^2} \, ,$$

we can write the following simple expression for $i \, \Gamma^{\rho\mu\nu}$

$$i \, \Gamma^{\rho\mu\nu} = -i \, \frac{e}{(4\pi)^2} \, \frac{M_W^2}{v^2} \, g^{\mu\nu} (k'^\rho - q^\rho) \, \frac{(-1)^\delta}{\delta} \int_0^1 dx \, \frac{(1 - 2x) \, x^\delta \, (1 - x)^\delta}{(k'^2 - M_x^2 - 2k' \cdot q)^{-\delta}}.$$

$$(8.104)$$

We can observe in the previous expression that we have a propagator-like denominator with a non-integer power given by "$-\delta$". Inserting this expression into the second loop, discarding terms proportional to γ_5 the expression for the vertex function Γ^μ reads

$$\Gamma^\mu = -i \, \frac{\alpha}{16 \, \pi \, s_w^2} \, \frac{M_W^2}{v^2} \, \frac{(-1)^\delta}{\delta} \int_0^1 dx \, (1 - 2x) \, x^\delta \, (1 - x)^\delta \, I^\mu, \qquad (8.105)$$

with $s_w = \sin\theta_w$ with θ_w the weak mixing angle[5] and with I^μ given by the integral

$$I^\mu = \int \frac{d^D k'}{(2\pi)^D} \frac{g^{\mu\nu}(k'^\rho - q^\rho)\,\gamma_\rho(\slashed{p} + \slashed{k}')\gamma_\nu}{(k'^2 - M_x^2 - 2k'\cdot q)^{-\delta}\,(k^2 - M_W^2)\,(p+k)^2\,[(k-q)^2 - M_W^2]}. \tag{8.106}$$

Particularizing (6.27) to our case we find

$$\frac{1}{A_1^{-\delta} A_2 A_3 A_4} = \frac{\Gamma(3-\delta)}{\Gamma(-\delta)} \int_0^1 dx_1 \int_0^{1-x_1} dx_2 \int_0^{1-x_1-x_2} dx_3 \, x_1^{-\delta-1}$$

$$\times \frac{1}{\left(x_1 A_1 + x_2 A_2 + x_3 A_3 + (1 - x_1 - x_2 - x_3) A_4\right)^{3-\delta}}$$

$$\equiv \frac{\Gamma(3-\delta)}{\Gamma(-\delta)} \int_0^1 dx_1 \int_0^{1-x_1} dx_2 \int_0^{1-x_1-x_2} dx_3 \, \frac{x_1^{-\delta-1}}{\mathcal{D}^{3-\delta}}. \tag{8.107}$$

Taking $A_1 = k'^2 - M_x^2 - 2k'\cdot q$, $A_2 = (k-q)^2 - M_W^2$, $A_3 = k^2 - M_W^2$ and $A_4 = (p+k)^2$ we obtain

$$\mathcal{D} = \left(k' - p(x_1 + x_2 + x_3 - 1) - q(x_1 + x_2)\right)^2 - b^2,$$

$$b^2 = M_x^2 x_1 + M_W^2(x_2 + x_3). \tag{8.108}$$

In the expression of b^2 we have ignored the muon mass as it is much smaller than all the other masses ($m \ll M_W, M_h$). With this Feynman parametrization, keeping only the $\sigma^{\mu\nu}q_\nu$ Lorentz structure[6] and the leading terms (linear in the muon mass) we obtain

$$I^\mu = i\,m\,\sigma^{\mu\nu}q_\nu \frac{\Gamma(3-\delta)}{\Gamma(-\delta)} \int_0^1 dx_1 \int_0^{1-x_1} dx_2 \int_0^{1-x_1-x_2} dx_3 \, x_1^{-\delta-1}\, x_3$$

$$\times \int \frac{d^D k'}{(2\pi)^D} \frac{1}{(k'^2 - b^2)^{3-\delta}}$$

$$= i\,m\,\sigma^{\mu\nu}q_\nu \frac{\Gamma(3-\delta)}{\Gamma(-\delta)} \int_0^1 dx_1 \int_0^{1-x_1} dx_2 \int_0^{1-x_1-x_2} dx_3 \, x_1^{-\delta-1}\, x_3$$

$$\times J(D, 0, 3 - \delta, b^2). \tag{8.109}$$

---

[5]Here we have used $g/2 = M_W/v$ and $e = g\,s_w$.

[6]In order to manipulate the spinor structures and write the final result as in (8.100) one must use the Dirac algebra introduced in Chap. 5 and the Gordon identity: $\bar{u}_r(p+q)(2p^\mu + q^\mu)u_s(p) = \bar{u}_r(p)(2m\gamma^\mu - i\sigma^{\mu\nu}q_\nu)u_s(p)$.

The function $J(D, 0, 3 - \delta, b^2)$ is finite for $D = 4$ and its expression is simply given by

$$
\begin{aligned}
J(4, 0, 3 - \delta, b^2) &= \frac{-i}{(4\pi)^2} (-1)^\delta (b^2)^{\delta - 1} \frac{\Gamma(1 - \delta)}{\Gamma(3 - \delta)} \\
&= \frac{-i}{(4\pi)^2} (-1)^\delta \frac{\Gamma(1 - \delta)}{\Gamma(3 - \delta)} \left[\frac{1}{b^2} + \frac{\delta}{b^2} \ln b^2 + O(\delta^2) \right]. \quad (8.110)
\end{aligned}
$$

Introducing the previous expression into (8.109) and integrating over x_1, x_2 and x_3, we obtain

$$
\begin{aligned}
I^\mu = \frac{\sigma^{\mu\nu} q_\nu}{2m} \frac{2m^2}{M_W^2} \frac{\Gamma(1 - \delta)}{\Gamma(-\delta)} \frac{(-1)^\delta}{(4\pi)^2} &\left(-\frac{1}{4\delta} \right. \\
&\left. + \frac{-3M_W^4 + 4M_W^2 M_x^2 - 2M_x^4 \ln(M_x^2/M_W^2) - M_x^4}{8(M_W^2 - M_x^2)^2} \right) \quad (8.111)
\end{aligned}
$$

Inserting this expression into (8.105), expanding in δ and taking the limit $\delta \to 0$ at the end, we finally find the contribution we were looking for

$$
\boxed{
\begin{aligned}
\Delta a_\mu = \frac{\alpha}{128\pi^3 s_w} \frac{m^2}{v^2} \int_0^1 dx\, (2x - 1) \\
\times \frac{3M_W^4 - 4M_W^2 M_x^2 + M_x^4 + 2M_x^4 \ln(M_x^2/M_W^2)}{8(M_W^2 - M_x^2)^2}
\end{aligned}
}
\quad (8.112)
$$

Further Reading

A. Pich, *Quantum Chromodynamics*, http://arxiv.org/pdf/hep-ph/9505231v1.pdf

V. Ilisie, *New Barr-Zee contributions to $(g - 2)_\mu$ in two-Higgs-doublet models.* JHEP **04**, 077 (2015), http://arxiv.org/pdf/1502.04199v3.pdf

M. Kaku, *Quantum Field Theory: A Modern Introduction*

M. Srednicki, *Quantum Field Theory*

S. Pokorsky, *Gauge Field Theories*

G. Dissertori, I.G. Knowles, M. Schmelling, *Quantum Chromodynamics.*

T. Muta, *Foundations of Quantum Chromodynamics*

M.E. Peskin, D.V. Schroeder, *An Introduction to Quantum Field Theory*

J.C. Romao, J.P. Silva, A resource for signs and Feynman diagrams of the standard model. Int. J. Mod. Phys. A **27**, 1230025 (2012), http://arxiv.org/pdf/1209.6213.pdf

L.H. Ryder, *Quantum Field Theory* (Cambridge University Press, Cambridge, 1985)

T.P. Cheng, L.F. Li, *Gauge Theory of Elementary Particle Physics* (Oxford University Press, Oxford, 1984)

F. Mandl, G.P. Shaw, *Quantum Field Theory*

Chapter 9
Massive Spin One and Renormalizable Gauges

Abstract For many decades of the last century, physicists were struggling to define consistent (renormalizable and unitarity preserving) models for spin-one massive particles (Proca fields). As we know, this was beautifully achieved by Weinberg, Salam and Glashow in 1967 when they proposed an electroweak unified theory which we now call the Standard Model. The electroweak symmetry breaking mechanism, among other things, generates mass terms for the W and Z bosons, while preserving renormalizability and unitarity. The longitudinal degrees of freedom of the massive spin-one particles are given by the Goldstone bosons. Choosing one gauge or another might seem just a matter of convenience and in most cases the unitary gauge is preferred. Here we will show, with an explicit example, that when performing loop calculations the unitary gauge is not really a good choice and that some Green functions are not renormalizable in this gauge. We will also show that working in the so-called renormalizable gauges completely fixes the problem. As it is a non-trivial task we shall also explicitly perform the renormalization of the W boson sector of the Standard Model and check the Ward identities with a simple one-loop example.

9.1 Unitary Gauge

In the unitary gauge, the Goldstone bosons are *eaten* by the longitudinal degrees of freedom of the W and Z bosons. The Lagrangian in this case is the most economical one, as we don't have the extra terms corresponding to the Goldstone fields. The number of needed diagrams for a given process (involving massive gauge bosons) reduces considerably when compared to other gauges. Given that all observables should be gauge invariant, choosing one gauge or another might just seem a matter of taste. However, when one tries to compute higher order loop calculations, one finds that some Green functions cannot be renormalized due to some special type of divergences (that we shall analyse in a moment with an example) that cannot be reabsorbed into the Z_i renormalization constants, as described in Chap. 7. One can argue that Green functions are not true observables and that when calculating

© Springer International Publishing Switzerland 2016

V. Ilisie, *Concepts in Quantum Field Theory*,

UNITEXT for Physics, DOI 10.1007/978-3-319-22966-9_9

Fig. 9.1 W boson contribution to the H self energy

$$i\,\Pi(q) \equiv$$

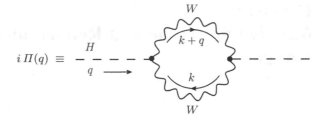

a true observable, the final result will render finite. This is completely true. However, calculations with divergent, non-renormalizable Green functions in the intermediate steps turns out to be a very difficult technical task.

Consider the W contribution to the *Higgs* self energy shown in Fig. 9.1. There are other diagrams (topologies) that involve W contributions but this is the one that is *problematic* and that we shall focus on. The needed interaction Lagrangian for this process is simply given by:

$$\mathcal{L}(x) = 2\frac{M_W^2}{v}H(x)W_\mu^\dagger(x)W^\mu(x). \tag{9.1}$$

We will also need the expression for W propagator. In the unitary gauge it has the canonical form of a massive spin-one particle:

$$iS^{\mu\nu}(k) = \frac{i}{k^2 - M_W^2}\left(-g^{\mu\nu} + \frac{k^\mu k^\nu}{M_W^2}\right). \tag{9.2}$$

With the previously given ingredients we can now calculate the $i\,\Pi(q)$ self-energy. Its explicit expression reads:

$$i\Pi(q^2) = \frac{4M_W^4}{v^2}\int\frac{d^D k}{(2\pi)^D}\left(-g_{\mu\nu} + \frac{k_\mu k_\nu}{M_W^2}\right)\left(-g^{\mu\nu} + \frac{(k+q)^\mu(k+q)^\nu}{M_W^2}\right)$$
$$\times\frac{1}{(k^2 - M_W^2)[(k+q)^2 - M_W^2]}$$
$$= \frac{4}{v^2}\int\frac{d^D k}{(2\pi)^D}\frac{(k^2 + k\cdot q)^2 - M_W^2(2k^2 + q^2 + 2k\cdot q) + DM_W^4}{(k^2 - M_W^2)[(k+q)^2 - M_W^2]}. \tag{9.3}$$

We are only interested in the UV divergent parts. Thus, (using the calculation tools introduced in the previous chapters) we obtain the following

$$\Pi_\epsilon(q) = \frac{-\mu^{2\epsilon}}{(4\pi v)^2}\frac{1}{\hat\epsilon}\left(12M_W^4 - 6q^2 M_W^2 + (q^2)^2\right), \tag{9.4}$$

where we have chosen the $\overline{MS}$ scheme for simplicity. We shall see that the $(q^2)^2$ term is the one that makes this Green function non-renormalizable.

Let's forget for a moment about this diagram and perform the Dyson summation for the Higgs propagator. Schematically it is given by

$$i S(q) =$$

where 1PI stands for *one particle irreducible*, as usual. Explicitly we have

$$i S(q) = i S^{(0)}(q) + i S^{(0)}(q) i \Pi(q) i S^{(0)}(q) + \dots$$

where $i S^{(0)}(q) = i/(q^2 - M_0^2)$ is the tree-level scalar boson propagator. The previous sum gives

$$S(q) = \frac{1}{q^2 - M_0^2} - \frac{\Pi(q)}{(q^2 - M_0^2)^2} + \dots = \frac{1}{q^2 - M_0^2 + \Pi(q)}, \qquad (9.5)$$

where M_0 is the bare mass of the Higgs field. Following the standard procedure for relating the non-renormalized propagator $S(q)$ with the renormalized one $S_R(q)$ (as described in Chap. 7), we define the renormalization constants as

$$S(q) = \frac{1}{q^2 - M_0^2 + \Pi(q)} \equiv Z_1 S_R(q) = \frac{Z_1}{q^2 - M^2 + \Pi_R(q)}, \qquad (9.6)$$

with M the renormalized Higgs mass given by

$$M^2 \equiv M_0^2 + \delta M^2. \qquad (9.7)$$

Parametrizing the Higgs self-energy as

$$\Pi(q) = (q^2 - M^2)\Pi_1 + \Pi_2, \qquad (9.8)$$

we are now able to calculate Z_1 and δM^2 explicitly:

$$
\begin{aligned}
Z_1 &= \frac{(q^2 - M^2)(1 + \Pi_{1,R}) + \Pi_{2,R}}{q^2 - M_0^2 + \Pi} = \frac{(q^2 - M^2)(1 + \Pi_{1,R}) + \Pi_{2,R}}{q^2 - M^2 + \delta M^2 + \Pi} \\
&= \frac{(q^2 - M^2)(1 + \Pi_{1,R}) + \Pi_{2,R}}{q^2 - M^2} \left(1 - \frac{\delta M^2 + \Pi}{q^2 - M^2}\right) \\
&= \left(1 + \Pi_{1,R} + \frac{\Pi_{2,R}}{q^2 - M^2}\right) \left(1 - \Pi_1 - \frac{\delta M^2 + \Pi_2}{q^2 - M^2}\right) \\
&= 1 - \Pi_{1,\epsilon} - \frac{\delta M^2 + \Pi_{2,\epsilon}}{q^2 - M^2}, \qquad (9.9)
\end{aligned}
$$

where $\Pi_{i,R}$ are regular parts of the self energy. Thus,

$$Z_1 = 1 - \Pi_{1,\epsilon}, \tag{9.10}$$

$$\delta M^2 = -\Pi_{2,\epsilon}. \tag{9.11}$$

Let's now go back to the W boson contribution to $\Pi(q)$ from (9.4). Parametrising this expression as in (9.8) we obtain

$$\Pi_{1,\epsilon} = \frac{-\mu^{2\epsilon}}{(4\pi v)^2} \frac{1}{\epsilon} \left(M^2 - 6 M_W^2 + q^2 \right), \tag{9.12}$$

$$\Pi_{2,\epsilon} = \frac{-\mu^{2\epsilon}}{(4\pi v)^2} \frac{1}{\epsilon} \left(M^4 - 6 M_W^2 M^2 + 12 M_W^4 \right). \tag{9.13}$$

It is obvious that the $\Pi_{1,\epsilon}$ term contains a q^2 term, which is forbidden. The renormalization constants cannot depend on any momenta. It is left for the reader as an exercise to analyse this case for the on-shell renormalization scheme.

9.2 R_ξ Gauges

The gauge fixing functional for the linear R_ξ gauges is given by

$$\mathcal{L}_{\text{GF}} = -\frac{1}{\xi} \left(\partial^\mu W_\mu^\dagger + i \xi M_W G^+ \right) \left(\partial^\mu W_\mu - i \xi M_W G^- \right), \tag{9.14}$$

where the W-Goldstone mixing term is *designed* to cancel the same mixing (but with opposite sign) generated by the covariant derivatives. Using this family of gauges is how 't Hooft proved the renormalizability of the Standard Model. In this gauge the W propagator reads

$$i S_\xi^{\mu\nu}(k) = \frac{i}{k^2 - M_W^2} \left(-g^{\mu\nu} + (1 - \xi) \frac{k^\mu k^\nu}{k^2 - \xi M_W^2} \right). \tag{9.15}$$

For $\xi \to \infty$ one recovers the unitary gauge expression (9.2). If one chooses, however, a finite value for this parameter, the Green functions of the theory are well-behaved.

The Lagrangian $\mathcal{L}_{\text{GF}}$ provides the Goldstone field with a mass term ξM_W^2. The Goldstone propagator is thus given by:

$$i S_{G\pm}(p) = \frac{i}{p^2 - \xi M_W^2}. \tag{9.16}$$

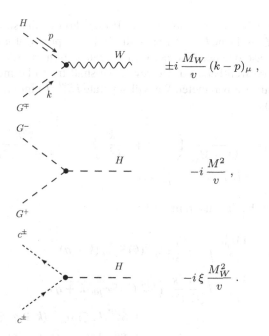

Fig. 9.2 Feynman rules for the $HW^\pm G^\mp$, HG^+G^- and $Hc^\pm \bar{c}^\pm$ interaction vertices

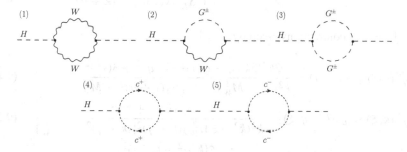

Fig. 9.3 Higgs self-energy diagrams involving Goldstone and W bosons and ghosts

In this gauge, we will also have Fadeev-Popov ghost ($c^\pm$) contributions associated to the gauge-fixing Lagrangian. The $c^\pm$ propagator has the same expression as the previous Goldstone propagator. Finally, the Feynman rules for the new cubic interactions involving one Higgs field are given in Fig. 9.2.

Including these new interactions, we now have four new diagrams corresponding to our analysed topology. The complete set is shown in Fig. 9.3. One should keep in mind that c^+ is not the anti-particle of c^-; their anti-particles are given by $\bar{c}^+$ and $\bar{c}^-$, thus both diagrams (4) and (5) must be taken into account (one should also remember to include a -1 sign corresponding to a closed ghost loop).

Looking at the diagrams form Fig. 9.3, the diagrams (1) and (2) are the only potentially dangerous ones, that can generate a $(q^2)^2/\hat{\epsilon}$ term. We shall see that none of them, however will generate such a term. If one chooses the so-called 't Hooft-Feynman gauge ($\xi \to 1$) the $k^\mu k^\nu$ terms of the W boson propagator are absent, thus in this particular case one needs to perform no calculation in order to see that the Green functions are well-behaved. However, we shall try to be more generic and work with an arbitrary ξ parameter. We will separate $i S_\xi^{\mu\nu}(k)$ into its transverse and a longitudinal parts

$$
i S_\xi^{\mu\nu}(k) = \frac{i}{k^2 - M_W^2}\left(-g^{\mu\nu} + \frac{k^\mu k^\nu}{k^2}\right) - i\xi \frac{k^\mu k^\nu}{k^2(k^2 - \xi M_W^2)}
$$
$$
\equiv i S_T^{\mu\nu}(k) + i S_L^{\mu\nu}(k, \xi).
\tag{9.17}
$$

The expression for the first diagram reads:

$$
i \Pi^{(1)}(q) = \frac{4 M_W^4}{v^2} \int \frac{d^D k}{(2\pi)^D} S_\xi^{\mu\nu}(k) S_{\xi,\mu\nu}(k + q)
$$
$$
= \frac{4 M_W^4}{v^2} \int \frac{d^D k}{(2\pi)^D} \Big(S_T^{\mu\nu}(k) S_{T,\mu\nu}(k + q)
$$
$$
+ S_L^{\mu\nu}(k, \xi) S_{L,\mu\nu}(k + q, \xi)
$$
$$
+ S_T^{\mu\nu}(k) S_{L,\mu\nu}(k + q, \xi)
$$
$$
+ S_L^{\mu\nu}(k, \xi) S_{T,\mu\nu}(k + q) \Big),
\tag{9.18}
$$

with the tensor contractions given by

$$
S_T^{\mu\nu}(k) S_{T,\mu\nu}(k + q) = \frac{2k^2(3k \cdot q + q^2) + (k \cdot q)^2 + 3(k^2)^2}{k^2(k^2 - M_W^2)(k + q)^2[(k + q)^2 - M_W^2]},
\tag{9.19}
$$

$$
S_L^{\mu\nu}(k, \xi) S_{L,\mu\nu}(k + q, \xi) = \frac{\xi^2(k \cdot q + k^2)^2}{k^2(k^2 - \xi M_W^2)(k + q)^2[(k + q)^2 - \xi M_W^2]},
\tag{9.20}
$$

$$
S_T^{\mu\nu}(k) S_{L,\mu\nu}(k + q, \xi) = \frac{\xi(k^2 q^2 - (k \cdot q)^2)}{k^2(k^2 - M_W^2)(k + q)^2[(k + q)^2 - \xi M_W^2]},
\tag{9.21}
$$

$$
S_L^{\mu\nu}(k, \xi) S_{T,\mu\nu}(k + q) = \frac{\xi(k^2 q^2 - (k \cdot q)^2)}{k^2(k^2 - \xi M_W^2)(k + q)^2[(k + q)^2 - M_W^2]}.
\tag{9.22}
$$

After introducing the Feynman parametrization, shifting the momentum in the denominator, etc., the only UV- divergent piece will be given by terms of the type[1]

$$
J(D, 2, 4, b^2) = J(D, 0, 2, b^2)\frac{D(D + 2)}{24}
$$
$$
= \frac{-i}{(4\pi)^2} \mu^{2\epsilon}\left[\frac{1}{\hat{\epsilon}} + \ln\left(\frac{a^2}{\mu^2}\right) + \frac{5}{6}\right] + \mathcal{O}(\epsilon),
\tag{9.23}
$$

---

[1] See Chap. 6 for details.

thus, no $(q^2)^2/\hat{\epsilon}$ is generated. The singular limit of the unitary gauge can be easily localized by taking $\xi \to \infty$ for the expressions (9.19)–(9.22). We observe that the tensor contraction of the longitudinal components of the two W boson propagators (9.20) in the $\xi \to \infty$ limit gives a contribution of the type

$$S_L^{\mu\nu}(k, \infty)S_{L,\mu\nu}(k + q, \infty) = \frac{1}{M_W^4} \frac{(k \cdot q + k^2)^2}{k^2(k + q)^2}, \qquad (9.24)$$

which is exactly the origin of our $(q^2)^2/\hat{\epsilon}$ singularity. With similar considerations one finds that diagram (2) is well-behaved.

Same type of divergences one will run into when considering for example W or Z self-energies. Thus, if a calculation involves (divergent) loops containing massive gauge bosons, the safest way of avoiding dangerous non-renormalizable terms is to simply avoid the unitary gauge. As mentioned before, the number of diagrams grows when Goldstone (and ghost) fields are included, however, these new diagrams are in most cases, trivial to compute.

For loops that give a finite final result i.e., $H \to \gamma\gamma$, one in principle has nothing to worry about. One will obtain the same finite result in any gauge. However, technically speaking, it is far more difficult to perform such a calculation in the unitary gauge even if the number of diagrams is significantly smaller.

9.3 Gauge Fixing Lagrangian and Renormalization

One natural question that may arise when renormalizing the massive W gauge boson[2] part of the Standard Model Lagrangian is, does the gauge fixing Lagrangian (9.14) need to be renormalized? There are two approaches. One can choose to renormalize the gauge fixing Lagrangian and reabsorb divergent parts into the gauge parameter as in (7.20) for the QED case, or, one can consider that (9.14) is already the gauge-fixing functional for the renormalized Lagrangian, also similar to the QED case (where we gave the alternative prescription to simply choose the Feynman gauge for the tree level propagator and ignore the $q^\mu q^\nu$ parts when summing higher order corrections, as they will never contribute to physical observables). The first approach is the most complicated one from a technical point of view as it would require two (and not one) bare gauge fixing parameters i.e., the gauge fixing functional in terms of bare quantities would be given by

$$\mathcal{L}_{GF} = -\frac{1}{\xi_1^{(0)}} \left(\partial^\mu W_\mu^{(0)\dagger} + i\xi_2^{(0)} M_W^{(0)} G_{(0)}^+\right)\left(\partial^\mu W_\mu^{(0)} - i\xi_2^{(0)} M_W^{(0)} G_{(0)}^-\right), \qquad (9.25)$$

---

[2]Here we shall only focus on the W gauge boson, however, one can extend the discussion to the Z boson.

and only after renormalizing, one can choose the renormalized gauge fixing parameters to be equal ($\xi_1 = \xi_2 = \xi$). Thus, the renormalization procedure is in this case highly complex. Here we shall present the alternative procedure.

We shall choose the Feynman gauge ($\xi = 1$) for this discussion.[3] Expanding the gauge fixing functional from (9.14) we obtain

$$\mathcal{L}_{\text{GF}} = -\partial^\mu W_\mu^\dagger \partial^\nu W_\nu - M_W^2 G^+ G^- + i M_W \partial^\mu W_\mu^\dagger G^- - i M_W \partial^\mu W_\mu G^+. \quad (9.26)$$

The gauge-Goldstone mixture from the Standard Model Lagrangian (that comes from the covariant derivatives) in terms of the bare quantities reads

$$\mathcal{L} = -i M_W^{(0)} \partial^\mu W_\mu^{(0)\dagger} G_{(0)}^- + i M_W^{(0)} \partial^\mu W_\mu^{(0)} G_{(0)}^+. \quad (9.27)$$

Introducing the renormalization constants as

$$M_W^{(0)} = Z_M^{1/2} M_W, \qquad W_\mu^{(0)} = Z_W^{1/2} W_\mu,$$
$$G_{(0)}^\pm = Z_{G^\pm}^{1/2} G^\pm, \quad (9.28)$$

(where $Z_i = 1 + \delta_i$) the previous Lagrangian (9.27) in terms of the renormalized parameters reads

$$
\begin{aligned}
\mathcal{L} &= -i Z_W^{1/2} Z_M^{1/2} Z_{G^\pm}^{1/2} M_W \partial^\mu W_\mu^\dagger G^- + \text{h.c.} \\
&= -i \left(1 + \frac{1}{2}\delta_W + \frac{1}{2}\delta_M + \frac{1}{2}\delta_{G^\pm} \right) M_W \partial^\mu W_\mu^\dagger G^- + \text{h.c.} \\
&= -i M_W \partial^\mu W_\mu^\dagger G^- - \frac{i}{2}\left(\delta_W + \delta_M + \delta_{G^\pm} \right) M_W \partial^\mu W_\mu^\dagger G^- + \text{h.c.} \\
&= \tilde{\mathcal{L}} + \delta\mathcal{L}.
\end{aligned}
\quad (9.29)
$$

The tree-level gauge-Goldstone mixing given by $\tilde{\mathcal{L}}$ is cancelled by the same mixing from the gauge fixing Lagrangian (9.26) and $\delta\mathcal{L}$ given by

$$
\begin{aligned}
\delta\mathcal{L} &= -\frac{i}{2}\left(\delta_W + \delta_M + \delta_{G^\pm} \right) M_W \partial^\mu W_\mu^\dagger G^- + \text{h.c.} \\
&\equiv -i Z_{WG} M_W \partial^\mu W_\mu^\dagger G^- + \text{h.c.}
\end{aligned}
\quad (9.30)
$$

is a counterterm for the gauge-Goldstone mixing that appears at one-loop level. The Feynman rule for this counterterm reads[4]

---

[3] Any other finite choice for the value of ξ is perfectly valid, however this is the choice that mostly simplifies the gauge boson propagator.

[4] An useful trick for deriving the Feynman rules corresponding to interaction terms that contain derivatives of fields is given in Appendix C.

$$\mp i\, p_\mu\, M_W\, Z_{WG}$$

Thus, the following sum is finite

where the grey blobs from the LHS represent the sum of all the one-loop self-energy diagrams.

In the following we shall formally perform the Dyson resummation and renormalize the W boson propagator, as this is a special case and the procedure is not as straightforward as in the previously studied cases. The W boson self-energy can be written as

$$
i\Pi_W^{\mu\nu} \equiv ig^{\mu\nu}\Pi_W^T + i\frac{p^\mu p^\nu}{p^2}\Pi_W^L
$$

$$
\equiv i\left(g^{\mu\nu} - \frac{p^\mu p^\nu}{p^2}\right)\Pi_W^T + i\frac{p^\mu p^\nu}{p^2}\tilde{\Pi}_W^L \tag{9.31}
$$

with $\tilde{\Pi}_W^L = \Pi_W^L + \Pi_W^T$. Performing the Dyson resummation for the propagator in the Feynman gauge we have

$$
\begin{aligned}
iS^{\mu\nu} &= \frac{-ig^{\mu\nu}}{p^2 - M_W^{(0)2}} + \frac{-ig^{\mu\lambda}}{p^2 - M_W^{(0)2}}\left(i\Pi_{\lambda\rho}\right)\frac{-ig^{\rho\nu}}{p^2 - M_W^{(0)2}} \\
&\quad + \frac{-ig^{\mu\lambda}}{p^2 - M_W^{(0)2}}\left(i\Pi_{\lambda\rho}\right)\frac{-ig^{\rho\sigma}}{p^2 - M_W^{(0)2}}\left(i\Pi_{\sigma\delta}\right)\frac{-ig^{\delta\nu}}{p^2 - M_W^{(0)2}} + \cdots \\
&= \frac{-ig^{\mu\nu}}{p^2 - M_W^{(0)2}} + \frac{-i}{\left(p^2 - M_W^{(0)2}\right)^2}\left(g^{\mu\lambda}\Pi_{\lambda\rho}g^{\rho\nu}\right) \\
&\quad + \frac{-i}{\left(p^2 - M_W^{(0)2}\right)^3}\left(g^{\mu\lambda}\Pi_{\lambda\rho}g^{\rho\sigma}\Pi_{\sigma\delta}g^{\delta\nu}\right) + \cdots
\end{aligned}
$$

$$= \frac{-i g^{\mu\nu}}{p^2 - M_W^{(0)2}} \left(1 + \frac{\Pi_W^T}{p^2 - M_W^{(0)2}} + \frac{(\Pi_W^T)^2}{\left(p^2 - M_W^{(0)2}\right)^2} + \cdots \right)$$

$$+ \frac{-i p^\mu p^\nu}{p^2} \frac{\Pi_W^L}{\left(p^2 - M_W^{(0)2}\right)^2} \left(1 + \frac{\Pi_W^L + 2\Pi_W^T}{p^2 - M_W^{(0)2}} + \cdots \right). \quad (9.32)$$

Thus, we find

$$i S^{\mu\nu} = \frac{-i g^{\mu\nu}}{p^2 - M_W^{(0)2} - \Pi_W^T} + \frac{-i p^\mu p^\nu}{p^2} \times$$

$$\times \frac{\Pi_W^L}{\left(p^2 - M_W^{(0)2} - \Pi_W^T\right)\left(p^2 - M_W^{(0)2} - \Pi_W^T - \Pi_W^L\right)} \qquad (9.33)$$

The renormalized propagator is then given by

$$i S^{\mu\nu} = Z_W \frac{-i g^{\mu\nu}}{p^2 - M_W^2 - \Pi_{W,R}^T} + Z_W \frac{-i p^\mu p^\nu}{p^2} \times$$

$$\times \frac{\Pi_{W,R}^L}{\left(p^2 - M_W^2 - \Pi_{W,R}^T\right)\left(p^2 - M_W^2 - \Pi_{W,R}^T - \Pi_{W,R}^L\right)}$$

$$\equiv i Z_W S_R^{\mu\nu} \qquad (9.34)$$

Using the standard procedure, one can now relate the Z_W and Z_M renormalization constants with the self-energies at the one-loop level. We shall shortly see with an explicit example that the divergences of Π_W^T and Π_W^L are related and only one constant Z_W is needed for the wave function renormalization. Thus, it is worth mentioning that the mass and wave function counterterms are totally determined by Π_W^T and one can use only the $g^{\mu\nu}$ part of the propagator to calculate Z_W and Z_M. Starting at the two-loop level things are a little more complicated, as the previous Dyson summation must also include one-loop transitions of the type $W - G^\pm$:

However, two-loop renormalization is far beyond the goal of this book.

What about tadpoles? The widely spread approach (and the most simple) is to perform a redefinition of the Higgs field of the Standard Model that generates a counterterm that cancels the tadpole contributions to the self-energies, as we did for the ϕ^3 model in Chap. 7.[5] Thus, if adopting this scheme, one only has to deal with

---

[5]This corresponds to the β_h scheme from S. Actis, A. Ferroglia, M. Passera, G. Passarino, *Two-Loop Renormalization in the Standard Model. Part I: Prolegomena*, Nucl. Phys. B **777** (2007) 1, http://arxiv.org/pdf/hep-ph/0612122.pdf.

1PI diagrams for the previous calculations. However, among the 1PI diagrams, the Goldstone self-energies include indirectly a tadpole contribution as we shall see in a moment. After shifting the vacuum expectation value of H one obtains a counterterm Lagrangian

$$\mathcal{L}_{ct} = -\beta_h \left(\frac{2M_W^2}{g^2} + \frac{2M_W}{g} H + \frac{1}{2}(H^2 + G_0^2 + 2G^+G^-) \right), \qquad (9.35)$$

which, as we have already mentioned, cancel tadpole contributions in all self-energies as it is shown in Fig. 9.4, but indirectly adds $-\beta_h$ contributions to $G^\pm$, G^0 and H self energies as shown in Fig. 9.5. The β_h constant is shown in Fig. 9.6. Thus, as we can observe, the renormalization of the massive Gauge boson and Goldstone terms of the Standard Model Lagrangian is highly non-trivial.

It is also worth mentioning the following detail. The W and $G^\pm$ self-energies and $W - G^\pm$ mixing are not all independent. They are related through a Ward identity. Diagrammatically it (the doubly contracted Ward identity) translates into

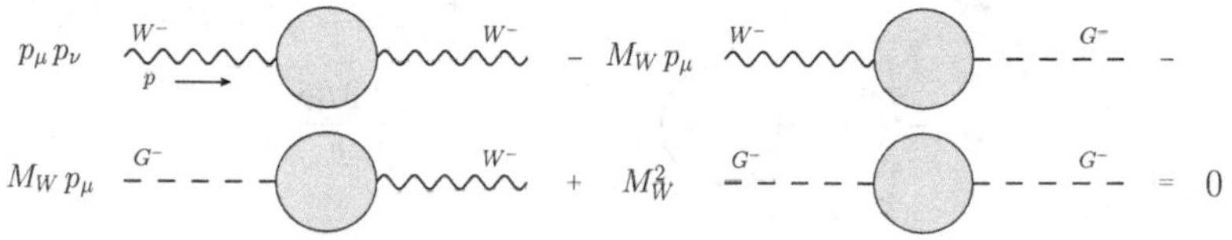

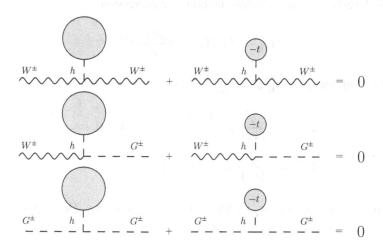

If we adopt the previous renormalization scheme (with no tadpole contributions except for the *indirect* one present in the Goldstone self-energy shown in Fig. 9.5)

Fig. 9.4 Examples of tadpole cancellation in the Standard Model

Fig. 9.5 Tadpole contribution to the 1PI Higgs and Goldstone self-energies

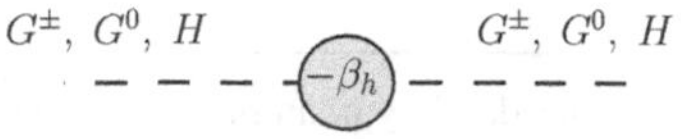

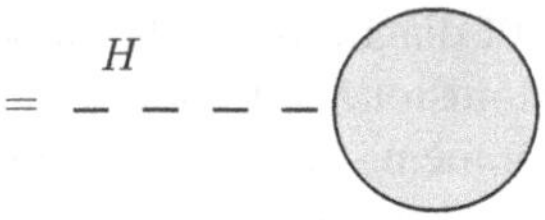

Fig. 9.6 Higgs tadpole contribution in the Standard Model and definition of β_h

$$it = i\beta_h \frac{2M_W}{g} = \quad \text{-- -- -- --} \bigcirc$$

one does not have to consider tadpoles either in the previous self-energies (except for the indirect ones)!

In order to give a little more sense to the previously presented abstract renormalization procedure and the Ward identity we shall take a simple one-loop explicit example. Consider the following one-loop lepton-neutrino contribution to the W and $G^\pm$ self-energies, $W - G^\pm$ mixing and tadpole[6]

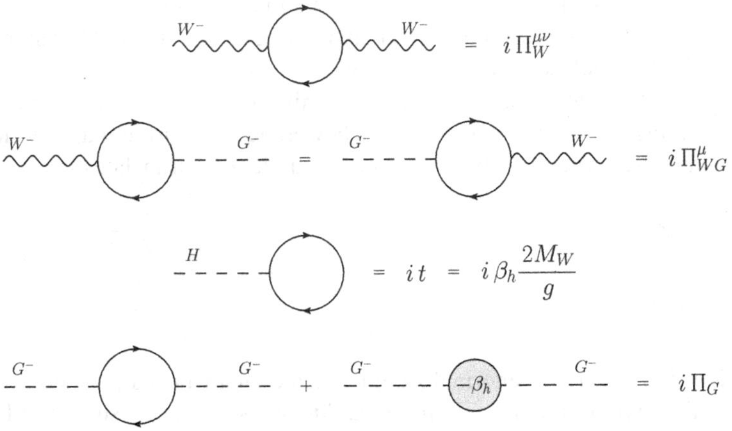

For the W self-energy one obtains the following expression

$$i\Pi_W^{\mu\nu} = ig^{\mu\nu}\Pi_W^T + i\frac{p^\mu p^\nu}{p^2}\Pi_W^L, \tag{9.36}$$

with the form functions Π_W^T and Π_W^L given by

$$\Pi_W^T = -\frac{g^2}{(4\pi)^2}\mu^{2\epsilon}\frac{1}{\hat{\epsilon}}\left(\frac{1}{2}m_l^2 - \frac{1}{3}p^2\right) + \text{finite}, \tag{9.37}$$

$$\Pi_W^L = -\frac{g^2}{(4\pi)^2}\mu^{2\epsilon}\frac{1}{\hat{\epsilon}}\left(\frac{1}{3}p^2\right) + \text{finite}, \tag{9.38}$$

where $a^2 = -p^2 x(1-x) + m_l^2 x$ with m_l the mass of the lepton. The tadpole diagram gives

$$t = \frac{4m_l^4}{(4\pi)^2 v}\mu^{2\epsilon}\frac{1}{\hat{\epsilon}} + \text{finite}. \tag{9.39}$$

---

[6]The needed Feynman rules are given at the end of the chapter.

with $\beta_h = gt/2M_W$. Summing the two corresponding diagrams for the Goldstone self-energy we get

$$\Pi_G = \frac{g^2}{(4\pi)^2} \frac{m_l^2}{M_W^2} \mu^{2\epsilon} \frac{1}{\hat{\epsilon}} \left(m_l^2 - \frac{1}{2}p^2 \right) - \beta_h + \text{finite}$$

$$= -\frac{g^2}{(4\pi)^2} \frac{m_l^2}{M_W^2} \mu^{2\epsilon} \frac{1}{\hat{\epsilon}} \left(\frac{1}{2}p^2 \right) + \text{finite}, \tag{9.40}$$

where we have used $g/2 = M_W/v$. The $W - G^\pm$ mixed term gives

$$\Pi_{WG}^\mu = -p^\mu \frac{g^2}{(4\pi)^2} \frac{m_l^2}{M_W} \mu^{2\epsilon} \frac{1}{\hat{\epsilon}} \frac{1}{2} + \text{finite}. \tag{9.41}$$

One can, thus, easily check that the following Ward identity holds

$$\boxed{p_\mu p_\nu \Pi_W^{\mu\nu} + M_W^2 \Pi_G - 2M_W p_\mu \Pi_{WG}^\mu = 0}. \tag{9.42}$$

Here we have checked this identity only for the UV-pole containing parts of the self-energies. It is left as an exercise for the reader to try to check this relation also for the finite parts. Let's also explicitly check that

$$\boxed{i\Pi_{WG}^\mu + iM_W p^\mu Z_{WG} = \text{finite}}. \tag{9.43}$$

Separating the W and $G^\pm$ self energies as:

$$\Pi_W^T = (p^2 - M_W^2)\Pi_{W,1}^T + \Pi_{W,2}^T, \tag{9.44}$$

$$\Pi_G = (p^2 - M_W^2)\Pi_{G,1} + \Pi_{G,2}, \tag{9.45}$$

we obtain the following counter-terms in the $\overline{MS}$ scheme

$$\delta_W = \Pi_{W,1}^T = \frac{1}{3} \frac{g^2}{(4\pi)^2} \mu^{2\epsilon} \frac{1}{\hat{\epsilon}}, \tag{9.46}$$

$$\delta_M = -\frac{\Pi_{W,2}^T}{M_W^2} = -\frac{g^2}{(4\pi)^2} \mu^{2\epsilon} \frac{1}{\hat{\epsilon}} \left(\frac{1}{3} - \frac{1}{2} \frac{m_l^2}{M_W^2} \right), \tag{9.47}$$

$$\delta_{G^\pm} = -\Pi_{G,1} = \frac{g^2}{(4\pi)^2} \mu^{2\epsilon} \frac{1}{\hat{\epsilon}} \left(\frac{1}{2} \frac{m_l^2}{M_W^2} \right), \tag{9.48}$$

where we have used the expressions from (9.28) for the Z_i constants and $Z_i = 1 + \delta_i$.

Thus

$$Z_{WG} \equiv \frac{1}{2}\left(\delta_M + \delta_W + \delta_{G^\pm}\right) = \frac{g^2}{(4\pi)^2}\mu^{2\epsilon}\frac{1}{\hat{\epsilon}}\left(\frac{1}{2}\frac{m_l^2}{M_W^2}\right), \qquad (9.49)$$

and (9.43) follows immediately. As we mentioned earlier, one must note that δ_W also cancels the infinities from Π_W^L (at one-loop level, when renormalizing the kinetic and the mass terms of the W boson one obtains a counterterm with the Feynman rule given by $i(p^\mu p^\nu - g^{\mu\nu}p^2) + ig^{\mu\nu}M_W^2(\delta_W + \delta_M)$ that cancels the poles of $i\Pi_W^{\mu\nu}$).

Finally one should notice that the mass counter-term obtained from the Goldstone boson propagator does not coincide with the one obtained from the W boson propagator. This is because the mass of the Goldstone is equal to the mass of the W boson, except, obviously, for quantum corrections which are different in the two cases. This should be even more obvious in the $m_l \to 0$ limit, for which the Goldstone couplings to fermions vanish (thus $\Pi_G = 0$), and as it can be observed in (9.47) one of the terms in δ_M still survives.

The needed Feynman rules for the fermionic loop calculations are shown below

$$-i\frac{g}{2\sqrt{2}}\frac{m_l}{M_W}(1-\gamma_5)$$

$$-i\frac{g}{2\sqrt{2}}\frac{m_l}{M_W}(1+\gamma_5)$$

$$-i\frac{g}{2\sqrt{2}}\gamma_\mu(1+\gamma_5)$$

$$-i\frac{m_l}{v}$$

Further Reading

K. Fujikawa, B.W. Lee, A.I. Sanda, Generalized renormalizable gauge formulation of spontaneously broken gauge theories. Phys. Rev. D **6**, 2923 (1972)

J.M. Cornwall, J. Papavassiliou, D. Binosi, *The Pinch Technique and its Applications to Non-Abelian Gauge Theories*. Cambridge Monographs on Particle Physics, Nuclear Physics and Cosmology

S. Actis, A. Ferroglia, M. Passera, G. Passarino, Two-loop renormalization in the standard model. part I: prolegomena. Nucl. Phys. B **777**, 1 (2007), http://arxiv.org/pdf/hep-ph/0612122.pdf

D. Bardin, G. Passarino, *The Standard Model in the Making, Precision Study of the Electroweak Interactions*, (Oxford University)Press

R. Santos, A. Barroso, On the renormalization of two Higgs doublet models. Phys. Rev. D **56**, 5366 (1997), http://arxiv.org/pdf/hep-ph/9701257.pdf

J.C. Romao, J.P. Silva, A resource for signs and Feynman diagrams of the Standard Model. Int. J. Mod. Phys. A **27**, 1230025 (2012), http://arxiv.org/pdf/1209.6213.pdf

V. Ilisie, *S.M. Higgs Decay and Production Channels*, http://ific.uv.es/lhcpheno/PhDthesis/master_vilisie.pdf

Chapter 10
Symmetries and Effective Vertices

Abstract When facing the computation of more realistic processes, the calculations can become lengthy very fast as the number of Feynman diagrams grows. Before starting the calculation process the problem should be reduced to its minimal form. Here we will present an example of how to reduce the number of calculated diagrams for a given process (which in this case will be a Higgs-like scalar decay to two photons through a charged scalar loop) using gauge symmetry.

10.1 Higgs Decay to a Pair of Photons

Consider the following interaction Lagrangian for scalar QED, with $H^{\pm}$ denoting the charged scalar, plus an hypothetical interaction between a Higgs-like neutral scalar ϕ and a pair of $H^{\pm}$

$$\mathcal{L} = i\, e\, A^{\mu}\, H^{+}\, \overleftrightarrow{\partial_{\mu}}\, H^{-} + e^2\, A^{\mu}\, A_{\mu}\, H^{+}\, H^{-} - v\, \lambda_{\phi H^{+} H^{-}}\, \phi\, H^{+}\, H^{-} \qquad (10.1)$$

where we have introduced the operator

$$a\, \overleftrightarrow{\partial_{\mu}}\, b \equiv a\, \partial_{\mu} b - (\partial_{\mu} a)\, b. \qquad (10.2)$$

This interaction Lagrangian is actually part of the Two-Higgs-Doublet Model (2HDM) extension of the Standard Model (SM) with v the electroweak symmetry breaking scale.[1] This extension assumes the presence of two Higgs doublets (instead of one, as in the SM), and it is characterized by three physical neutral scalars (instead of one) and a charged Higgs ($H^{\pm}$). In the previous interaction Lagrangian ϕ is assumed to be one of the neutral scalars of the model. The most important feature

---

[1] For a nice review of the SM read A. Pich, *The Standard Model of Electroweak Interactions,* http://arxiv.org/pdf/1201.0537.pdf. Also, for a nice review of the 2HDM read G.C. Branco, P.M. Ferreira, L. Lavoura, M.N. Rebelo, M. Sher and J.P. Silva, *Theory and phenomenology of two-Higgs-doublet models*, Phys. Rept. **516** (2012) 1, http://arxiv.org/pdf/1106.0034v3.pdf.

© Springer International Publishing Switzerland 2016
V. Ilisie, *Concepts in Quantum Field Theory*,
UNITEXT for Physics, DOI 10.1007/978-3-319-22966-9_10

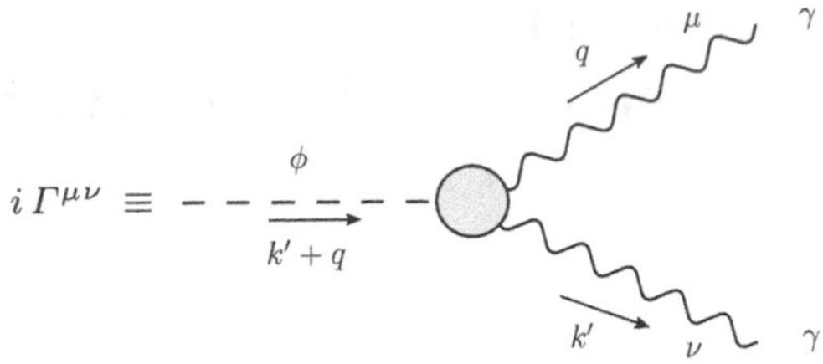

Fig. 10.1 One-loop $\phi \to \gamma\gamma$
effective vertex

of the 2HDM Lagrangian that we wish to exploit in this chapter, is that it preserves
gauge invariance (just like the SM).

Let's imagine that, given the previous interaction Lagrangian, we wish to calculate
the decay of the scalar ϕ to a pair of photons $\phi \to \gamma\gamma$. This decay starts at the
one loop level, as there are no tree-level interaction terms to account for it. Thus,
we can conclude that the result is finite and needs no renormalization (the 2HDM
is renormalizable). Before starting our calculation we can further exploit, as we
mentioned, gauge invariance, in order to simplify our computation. We can write
down an *effective vertex* describing the Lorentz structure of the final result. Consider
the following kinematic distribution for our process

$$\phi(k' + q) \to \gamma(q, \mu) \, \gamma(k', \nu) \tag{10.3}$$

(with $q^2 = k'^2 = 0$ and $(k' + q)^2 = M_\phi^2$ and with μ, ν Lorentz indices) as shown in
Fig. 10.1. The most generic Lorentz structure that this effective vertex can contain is

$$\begin{aligned}
\Gamma^{\mu\nu} = S \, g^{\mu\nu} \, k' \cdot q + A \, q^\mu q^\nu + B \, k'^\mu k'^\nu + C \, q^\mu k'^\nu \\
+ D \, q^\nu k'^\mu + E \, \epsilon^{\mu\nu\alpha\beta} k'_\alpha q_\beta,
\end{aligned} \tag{10.4}$$

where S, A, B, ..., E are Lorentz scalars. The last structure ($\sim \epsilon^{\mu\nu\alpha\beta}$) would imply
the presence of couplings proportional to γ_5 matrices, thus we can safely discard it.
Due to gauge invariance our effective vertex must satisfy

$$q_\mu \, \Gamma^{\mu\nu} = q_\nu \, \Gamma^{\mu\nu} = 0. \tag{10.5}$$

Also, all structures proportional to k'^ν and q^μ vanish when contacted with the polarization vectors of the photons (when calculating the transition matrix). Thus, one
obtains the following compact form for the effective vertex

$$\boxed{i \, \Gamma^{\mu\nu} = i \, (g^{\mu\nu} \, k' \cdot q - q^\nu k'^\mu) \, S} \, . \tag{10.6}$$

The transition matrix and its hermitian conjugate are thus given by

$$\begin{aligned}
-i \, \mathcal{M} &= i \, (g^{\mu\nu} k' \cdot q - k'^\mu q^\nu) \, S \, \epsilon_\mu^{r*}(q) \, \epsilon_\nu^{s*}(k') \\
i \, \mathcal{M}^\dagger &= -i \, (g^{\alpha\beta} k' \cdot q - k'^\alpha q^\beta) \, S^* \, \epsilon_\alpha^r(q) \, \epsilon_\beta^s(k').
\end{aligned} \tag{10.7}$$

Summing over the photon polarizations, making the usual substitution

$$\sum_r \epsilon_\mu^{r*}(p)\epsilon_\nu^r(p) \;\rightarrow\; -g_{\mu\nu}$$

(there is no triple gauge vertex in our present calculation[2] thus we need not to worry about ghosts) we find

$$\sum_{r,s} |\mathcal{M}|^2 = |S|^2 \, (g_{\mu\nu}k'\cdot q - k'_\mu q_\nu)(g^{\mu\nu}k'\cdot q - k'^\mu q^\nu) = \frac{1}{2}|S|^2 \, M_\phi^4. \qquad (10.8)$$

The phase space integral is trivial $\int dQ_2 = 1/(16\pi)$, thus we finally obtain

$$\boxed{\; \Gamma(H \rightarrow \gamma\gamma) \;=\; \frac{M_\phi^3}{64\pi} \, |S|^2 \;} \qquad (10.9)$$

The only remaining thing is to explicitly calculate the scalar form factor S.

Let's now focus on the following detail. We have proven, using symmetry arguments that our final result will have the form (10.6). Therefore, in order to calculate S we only need to keep the terms proportional to $k'^\mu q^\nu$ and ignore all the other Lorentz structures.[3] The Feynman diagrams that we need to compute are given by

Considering the distribution of loop momenta from Fig. 10.2, using the Feynman rules[4] given in Fig. 10.3 we obtain the following result for the first diagram

$$i\,\Gamma_{(1)}^{\mu\nu} = e^2 \, v^2 \, \lambda_{\phi H^+ H^-} \int_0^1 dx \int_0^1 dy \int \frac{d^D k}{(2\pi)^D} \frac{2x}{[k^2 - a^2]^3} \Big[4k^\mu k^\nu +$$
$$+ \, 2k'^\mu k'^\nu(2x^2 - 3x + 1) - k'^\mu q^\nu \, 4x(x-1)(y-1) \Big], \qquad (10.10)$$

where $a^2 = M_{H^\pm}^2 - M_\phi^2 x(1-x)(1-y)$. It is left for the reader as an exercise to show that the second diagram gives exactly the same result.

It is obvious from the Feynman rules that the third diagram only brings contributions to the $g^{\mu\nu}$ structure. Ignoring thus, this third diagram, keeping only the $k'^\mu q^\nu$ structure, we obtain

---

[2]Read Chap. 5 for more details.

[3]One could equally choose to keep only the $g^{\mu\nu} \, k'\cdot q$ structure, however, with this choice, one has to compute a greater number of terms and Feynman diagrams as we will shortly see.

[4]An useful trick for deriving the Feynman rules corresponding to interaction terms that contain derivatives of fields is given in Appendix C.

Fig. 10.2 Contribution of
the first diagram to the
$\phi \rightarrow \gamma\gamma$ process

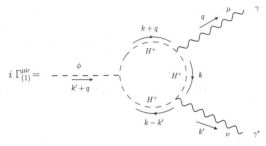

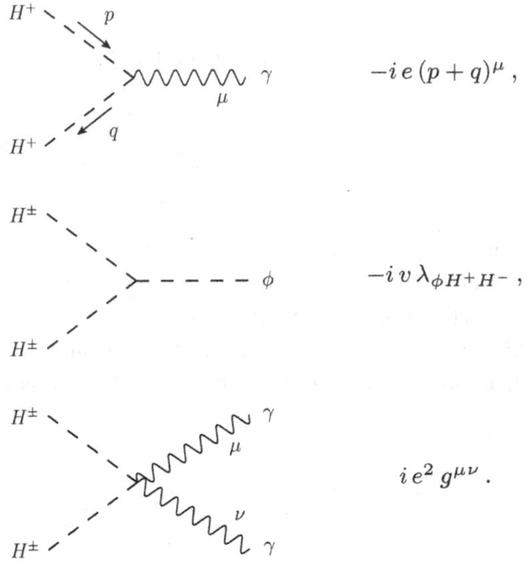

$i \, \Gamma^{\mu\nu}_{(1)} =$

Fig. 10.3 Feynman rules for
the interaction Lagrangian

$-i \, e \, (p + q)^{\mu} \, ,$

$-i \, v \, \lambda_{\phi H^+ H^-} \, ,$

$i \, e^2 \, g^{\mu\nu} \, .$

$$i \, \Gamma^{\mu\nu}_{(1+2)} = -e^2 v^2 \, \lambda_{\phi H^+ H^-} \int_0^1 dx \int_0^1 dy \int \frac{d^D k}{(2\pi)^D} \frac{16 x^2 (x-1)(y-1)}{[k^2 - a^2]^3} k'^{\mu} q^{\nu}$$

$$= i \, e^2 v^2 \, \lambda_{\phi H^+ H^-} \int_0^1 dx \int_0^1 dy \, \frac{16 x^2 (x-1)(y-1)}{32 \, \pi^2 \, a^2} k'^{\mu} q^{\nu} \qquad (10.11)$$

Thus, the scalar form factor S is simply given by

$$\boxed{S \; = \; -e^2 \, v^2 \, \lambda_{\phi H^+ H^-} \int_0^1 dx \int_0^1 dy \, \frac{x^2 (x-1)(y-1)}{2 \, \pi^2 \, a^2}} \, . \qquad (10.12)$$

Again, the reader is invited to explicitly check that this is exactly the result one
obtains by doing the full calculation.

In this example, using the symmetries of our Lagrangian, we were able to get
rid of one Feynman diagram and all the other Lorentz structures except $k'^{\mu} q^{\nu}$.

The UV-divergent parts of the structure $k^\mu k^\nu$ obviously cancels against the UV-divergent parts of the third diagram (that we have ignored).

It might seem at first sight that we haven't gained that much. This was, however a very simple example. If one has to calculate the full SM contribution to this decay (which involves fermions and W bosons) this technique tremendously simplifies the calculations. Other interesting examples can be found in the first two references.

Further Reading

V. Ilisie, A. Pich, Low-mass fermiophobic charged Higgs phenomenology in two-Higgs-doublet models. JHEP **1409**, 089 (2014). http://arxiv.org/pdf/1405.6639v3.pdf

V. Ilisie, New Barr-Zee contributions to $(\mathbf{g} - \mathbf{2})_\mu$ in two-Higgs-doublet models. JHEP **04**, 077 (2015) http://arxiv.org/pdf/1502.04199v3.pdf

A. Pich, *The Standard Model of Electroweak Interactions*, http://arxiv.org/pdf/1201.0537.pdf

G.C. Branco, P.M. Ferreira, L. Lavoura, M.N. Rebelo, M. Sher, J.P. Silva, Theory and phenomenology of two-Higgs-doublet models. Phys. Rep. **516**, 1 (2012) http://arxiv.org/pdf/1106.0034v3.pdf

C. Itzykson, J. Zuber, *Quantum Field Theory*

M.E. Peskin and D.V. Schroeder, *An Introduction to Quantum Field Theory*

Chapter 11
Effective Field Theory

Abstract Effective field theories (EFTs) are a highly important topic in Quantum Field Theory. Here we are going to shortly present some important highlights as well as the renormalization group equations for the Wilson coefficients. Afterwards we shall focus on one illustrative example and present the *matching* procedure at the one-loop level. The infrared behaviour of EFTs is also covered with this example.

11.1 Effective Lagrangian

For a nice introduction as well as more advanced topics consult references. An effective field theory is characterized by some effective Lagrangian:

$$\mathcal{L}_{eff} = \sum_i \frac{c_i}{\Lambda^{d_i-4}} \mathcal{O}_i \equiv \sum_i \mathcal{C}_i \mathcal{O}_i \,, \tag{11.1}$$

where $\mathcal{O}_i$ are operators constructed with the light fields, and the information on the heavy degrees of freedom is hidden in the couplings $\mathcal{C}_i$; d_i is the dimension of the operator $\mathcal{O}_i$ and Λ is the scale where the heavy fields come into the game. For simplicity we have considered that there is only one operator for a given dimension. If this was not the case, another summation (over all the operators of a given dimension) should be included in the Lagrangian i.e.,

$$\mathcal{L}_{eff} = \sum_{i,j} \frac{c_i^{(j)}}{\Lambda^{d_i-4}} \mathcal{O}_i^{(j)} \equiv \sum_{i,j} \mathcal{C}_i^{(j)} \mathcal{O}_i^{(j)} \,. \tag{11.2}$$

The operators are classified depending on their dimension as:

$$d_i < 4 \rightarrow \text{relevant}, \quad d_i = 4 \rightarrow \text{marginal}, \quad d_i > 4 \rightarrow \text{irrelevant}. \tag{11.3}$$

The Lagrangian $\mathcal{L}_{eff}$ contains an infinite number of terms (power series in Λ), therefore it is not renormalizable in the usual sense (an infinite number of counterterms

© Springer International Publishing Switzerland 2016   163
V. Ilisie, *Concepts in Quantum Field Theory*,
UNITEXT for Physics, DOI 10.1007/978-3-319-22966-9_11

would also be needed for renormalization). If we truncate the series of the Lagrangian i.e., we only include terms up to dimension N

$$\mathcal{L}_{eff} = \sum_i^N \mathcal{C}_i \mathcal{O}_i, \tag{11.4}$$

with N finite, the Lagrangian is not, in general, renormalizable either. However, for a given loop order the needed number of renormalization counterterms is finite.[1] In effective theories the number of counterterms needed for renormalization grows with the number of loop corrections included, whereas, for renormalizable theories (in the usual sense) the number of conterterms is the same at any loop order.

11.2 Renormalization Group Equations

11.2.1 No Operator Mixing

In the following we shall consider the renormalization group equations (RGEs) for the coefficients $\mathcal{C}_i$ and the operators $\mathcal{O}_i$. For this first analysis we shall only consider a simple case for which the coefficients do not mix under renormalization. Normally this is case when there is only one operator for a given dimension d_i and is what we are going to consider next. Given the bare operator $\mathcal{O}_i^B$ and the bare Wilson coefficient $\mathcal{C}_i^B$, when including loop corrections the Lagrangian can be written in terms of the renormalized quantities as:

$$\boxed{\mathcal{L}_{eff} = \sum_i \mathcal{C}_i^B \mathcal{O}_i^B = \sum_i \mathcal{C}_i^B Z_i(\mu)\mathcal{O}_i(\mu) = \sum_i \mathcal{C}_i(\mu)\mathcal{O}_i(\mu)}, \tag{11.5}$$

where we have defined the Z_i renormalization constant as

$$\mathcal{O}_i^B = Z_i(\mu)\mathcal{O}_i(\mu), \tag{11.6}$$

and where the renormalized Wilson coefficient is given by

$$\mathcal{C}_i^B = Z_i(\mu)^{-1}\mathcal{C}_i(\mu). \tag{11.7}$$

This is the standard approach when renormalizing effective Lagrangians; one includes the loop corrections, and hence the μ-dependence, that comes from the

---

[1]This is true provided that we use mass-independent renormalization schemes i.e., the $\overline{MS}$ scheme. In order to understand why mass-independent schemes are needed for renormalization of EFTs consult for example section **3. Quantum Loops** of *Effective Field Theory* (A. Pich, http://arxiv.org/abs/hep-ph/9806303).

operator renormalization into the Wilson coefficient so that the renormalized Lagrangian is μ-independent. All the loop corrections are then contained in the C_i coefficients (as we shall shortly see with an example). Let us write down the formal RGEs[2] for C_i and $\mathcal{O}_i$. The bare operator is μ-independent, therefore:

$$\mu \frac{d\mathcal{O}_i^B}{d\mu} = 0 = \mu\mathcal{O}_i \frac{dZ_i}{d\mu} + \mu Z_i \frac{d\mathcal{O}_i}{d\mu}. \tag{11.8}$$

We define the gamma function of the operator $\mathcal{O}_i$ as

$$\gamma_{\mathcal{O}_i} = \frac{\mu}{Z_i} \frac{dZ_i}{d\mu} = \gamma_{\mathcal{O}_i}^{(1)}\left(\frac{\alpha}{\pi}\right) + \gamma_{\mathcal{O}_i}^{(2)}\left(\frac{\alpha}{\pi}\right)^2 + \dots, \tag{11.9}$$

where α is the expansion parameter of the operator. Thus, the RGE for the operator $\mathcal{O}_i$ is simply given by:

$$\mu \frac{d\mathcal{O}_i}{d\mu} = -\gamma_{\mathcal{O}_i} \mathcal{O}_i. \tag{11.10}$$

On the other hand, the product $\mathcal{O}_i^B C_i^B = \mathcal{O}_i(\mu)\mathcal{C}(\mu)$ is μ-independent, thus we obtain analogous RGEs for the Wilson coefficients C_l:

$$\boxed{\mu \frac{dC_i}{d\mu} = \gamma_{\mathcal{O}_i} C_i}. \tag{11.11}$$

The solution of this equation is given in (B.23):

$$\boxed{\begin{aligned} C_i(\mu) &= C_i(\mu_0) \exp\left\{\int_{\alpha(\mu_0)}^{\alpha(\mu)} \frac{d\alpha}{\alpha} \frac{\gamma_{\mathcal{O}_i}(\alpha)}{\beta(\alpha)}\right\} \\ &= C_i(\mu_0)\left[\frac{\alpha(\mu)}{\alpha(\mu_0)}\right]^{\gamma_{\mathcal{O}_i}^{(1)}/\beta_1}\left\{1 + \dots\right\} \end{aligned}} \tag{11.12}$$

11.2.2 Operator Mixing

Let's now consider that there are n operators for a given dimension d_i. Introducing the short-hand notation $\mathcal{O}_i^{(j)} \equiv \mathcal{O}_j$ (with $j = 1, \dots, n$), the effective Lagrangian for the given dimension d_i can be written as

---

[2]For a short reminder of RGEs see Appendix B.

$$\mathcal{L}_{i,\text{eff}} = \sum_j \mathcal{C}_i^{(j)} \mathcal{O}_i^{(j)} \equiv \sum_j \mathcal{C}_j \mathcal{O}_j \equiv \mathcal{C}^T \mathcal{O}, \tag{11.13}$$

where we have defined $\mathcal{O}$ as a 1-column vector and $\mathcal{C}^T$ as a 1-row vector. In general the operators mix under renormalization, i.e.,

$$\mathcal{O}_i^B = \sum_j^n Z_{ij}(\mu)\mathcal{O}_j(\mu) \quad \Rightarrow \quad \mathcal{C}_i^B = \sum_j^n \left(Z^{-1}\right)_{ji}(\mu)\mathcal{C}_j(\mu). \tag{11.14}$$

In this case it turns out to be very useful to use the matrix notation:

$$\mathcal{O}^B = \mathbf{Z}(\mu)\mathcal{O}(\mu) \quad \Rightarrow \quad \mathcal{C}^B = \left(\mathbf{Z}^{-1}\right)^T(\mu)\mathcal{C}(\mu). \tag{11.15}$$

We easily obtain the RGEs for the operators and the Wilson coefficients

$$\boxed{\left(\mu\frac{d}{d\mu} + \gamma_{\mathcal{O}}\right)\mathcal{O}(\mu) = 0 , \qquad \left(\mu\frac{d}{d\mu} - \gamma_{\mathcal{O}}^T\right)\mathcal{C}(\mu) = 0}$$
$$\tag{11.16}$$

where the anomalous dimension matrix $\gamma_{\mathcal{O}}$ is given by:

$$\gamma_{\mathcal{O}} \equiv \mathbf{Z}^{-1}\mu\frac{d}{d\mu}\mathbf{Z}. \tag{11.17}$$

In order to solve this equation we must find the matrix $\mathbf{U}$ that diagonalizes $\gamma_{\mathcal{O}}$ i.e.,

$$\mathbf{U}^{-1}\gamma_{\mathcal{O}}^T\mathbf{U} = \tilde{\gamma}_{\mathcal{O}}, \tag{11.18}$$

with $\tilde{\gamma}_{\mathcal{O}}$ diagonal. Introducing the previous equation into (11.16) we find:

$$\left(\mathbf{U}\mathbf{U}^{-1}\mu\frac{d}{d\mu} - \mathbf{U}\tilde{\gamma}_{\mathcal{O}}\mathbf{U}^{-1}\right)\mathcal{C}(\mu) = 0. \tag{11.19}$$

Defining $\tilde{\mathcal{C}} = \mathbf{U}^{-1}\mathcal{C}$ we find the following equations in diagonal form:

$$\left(\mu\frac{d}{d\mu} - \tilde{\gamma}_{\mathcal{O}}\right)\tilde{\mathcal{C}}(\mu) = 0 \tag{11.20}$$

Thus, the coefficients $\tilde{\mathcal{C}}_i$ obey the unmixed RGEs:

$$\left(\mu\frac{d}{d\mu} - \tilde{\gamma}_{\mathcal{O}}^{(i)}\right)\tilde{\mathcal{C}}_i(\mu) = 0, \tag{11.21}$$

where $\tilde{\gamma}_{\mathcal{O}}^{(i)}$ are the diagonal terms of the matrix $\tilde{\gamma}_{\mathcal{O}}$. The solution is straightforward (same as in the previous section):

$$\tilde{C}_i(\mu) = \tilde{C}_i(\mu_0) \exp \left\{ \int_{\alpha(\mu_0)}^{\alpha(\mu)} \frac{d\alpha}{\alpha} \frac{\tilde{\gamma}_{\mathcal{O}}^{(i)}(\alpha)}{\beta(\alpha)} \right\}. \tag{11.22}$$

Changing the basis from $\tilde{C}_i$ to C_i we finally find the solution we were looking for:

$$\boxed{C_i(\mu) = \sum_{j,k} \mathbf{U}_{ij} \exp \left\{ \int_{\alpha(\mu_0)}^{\alpha(\mu)} \frac{d\alpha}{\alpha} \frac{\tilde{\gamma}_{\mathcal{O}}^{(j)}(\alpha)}{\beta(\alpha)} \right\} (\mathbf{U}^{-1})_{jk} C_k(\mu_0).} \tag{11.23}$$

11.3 Matching

11.3.1 $\varphi\phi^2$ Theory

Consider the following Lagrangian:

$$\mathcal{L} = \frac{1}{2}(\partial\phi)^2 - \frac{1}{2}m^2\phi^2 + \frac{1}{2}(\partial\varphi)^2 - \frac{1}{2}M^2\varphi^2 - \frac{1}{2}\lambda\varphi\phi^2 \tag{11.24}$$

If $M \gg m, E$, where E is the energy domain of the process we want to analyse, we can integrate out (from the path integral) the heavy field φ and write down an effective Lagrangian that describes physical processes at energies $E \sim m$. Say we wish to describe a scattering process like $\phi(p_1)\phi(p_2) \to \phi(p_3)\phi(p_4)$, then we can write the following effective Lagrangian:

$$\mathcal{L}_{eff} = \frac{1}{2}\bar{C}_0(\partial\phi)^2 - \frac{1}{2}C_2\phi^2 - \frac{1}{4!}C_0\phi^4 + \dots, \tag{11.25}$$

where $\bar{C}_0, C_0$ are the Wilson coefficients of zero dimension and C_2 is the two-dimensional one. These coefficients can be expanded as:

$$\bar{C}_0 = \bar{C}_0^{(0)} + \bar{C}_0^{(1)} + \dots, \quad C_0 = C_0^{(0)} + C_0^{(1)} + \dots, \quad C_2 = C_2^{(0)} + C_2^{(1)} + \dots, \tag{11.26}$$

where the upper index stands for the corresponding loop order. In the following we shall present the matching procedure. Both theories must give identical results at a matching scale $\mu \sim M$. Thus at this given scale the scattering amplitudes provided by $\mathcal{L}$ and $\mathcal{L}_{eff}$ must be the same. At tree-level however, the matching is scale independent (the scale dependence comes into the game only at the loop level). The tree-level matching is diagrammatically shown in Fig. 11.1. Explicitly this reads:

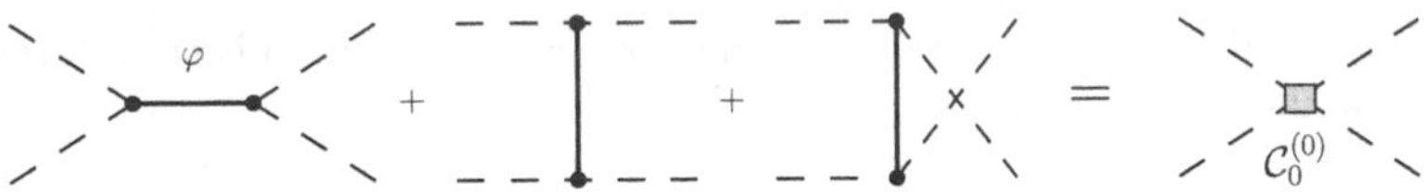

Fig. 11.1 Tree-level matching conditions

$$\boxed{\mathcal{M}^{(0)} = \mathcal{M}_{eff}^{(0)}}. \tag{11.27}$$

Dropping the $-i$ global factor from $-i\mathcal{M}$, the amplitude of the full theory for the scattering process is given by

$$\mathcal{M}^{(0)} = -i\lambda^2 \left(\frac{1}{s^2 - M^2} + \frac{1}{t^2 - M^2} + \frac{1}{u^2 - M^2} \right)$$

$$= i3\frac{\lambda^2}{M^2} + O(M^{-4}). \tag{11.28}$$

For the effective theory the amplitude simply reads

$$\mathcal{M}_{eff}^{(0)} = -i\mathcal{C}_0^{(0)}. \tag{11.29}$$

Comparing the two results we obtain the tree-level matching conditions

$$\mathcal{C}_0^{(0)} = -3\frac{\lambda^2}{M^2}. \tag{11.30}$$

It obvious that at this level $\bar{\mathcal{C}}_0^{(0)} = 1$ and $\mathcal{C}_2^{(0)} = m^2$ by comparing the propagators of the light scalar ϕ in both theories.

At one-loop level things start to be a little more difficult. The needed diagrams are shown in Fig. 11.2.

In this analysis we shall only compute the first set of diagrams (first line of Fig. 11.2). The second line is left as an exercise for the reader. The first two diagrams of the complete theory give the following contribution to the scattering amplitude:

$$\mathcal{M}^{(1,a)} = i\frac{1}{2}\frac{\lambda^2}{(4\pi)^2}\frac{m^2}{M^2}\mu^{2\epsilon}\left[\frac{1}{\hat{\epsilon}} + \ln\left(\frac{m^2}{\mu^2}\right) - 1\right],$$

$$\mathcal{M}^{(1,b)} = -i\frac{\lambda^2}{(4\pi)^2}\mu^{2\epsilon}\left[\frac{1}{\hat{\epsilon}} + \int_0^1 dx \ln\left(\frac{a^2}{\mu^2}\right)\right], \tag{11.31}$$

where $a^2 = p^2 x(x-1) + m^2 x + M^2(1-x)$. After renormalizing in the $\overline{MS}$ scheme we find

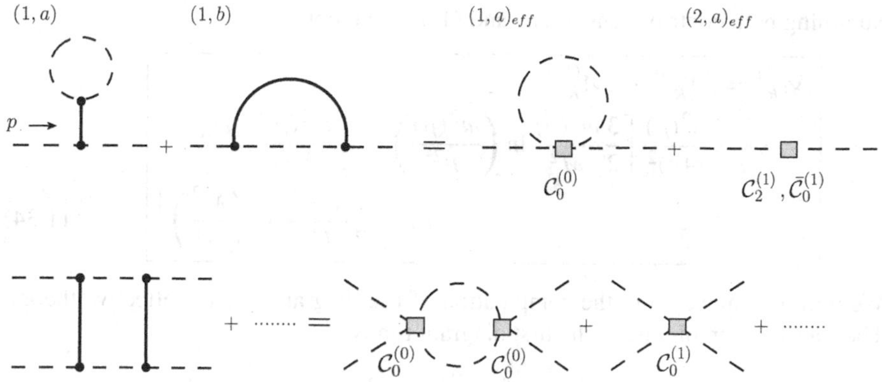
(1, a)
(1, b)
(1, a)eff
(2, a)eff
p
$C_0^{(0)}$
$C_2^{(1)}, \bar{C}_0^{(1)}$
$C_0^{(0)}$
$C_0^{(0)}$
$C_0^{(1)}$

Fig. 11.2 One-loop matching conditions

$$
\begin{aligned}
\mathcal{M}_R^{(1,a)} &= i\frac{1}{2}\frac{\lambda^2(\mu)}{(4\pi)^2}\frac{m^2(\mu)}{M^2}\left[\ln\left(\frac{m^2(\mu)}{\mu^2}\right)-1\right], \\
\mathcal{M}_R^{(1,b)} &= -i\frac{\lambda^2(\mu)}{(4\pi)^2}\int_0^1 \ln\left(\frac{a^2(\mu)}{\mu^2}\right),
\end{aligned}
\tag{11.32}
$$

where $\lambda(\mu)$ and $m(\mu)$ are the running coupling and the running mass of the full theory (as usual, the subscript R stands for renormalized or regular).[3] In order to perform the matching properly we must expand the logarithm of a^2/μ^2 in powers of p^2/M^2:

$$
\begin{aligned}
\int_0^1 dx \ln\left(\frac{a^2}{\mu^2}\right) &= \int_0^1 dx \ln\left(\frac{p^2x(x-1)+m^2x+M^2(1-x)}{\mu^2}\right) \\
&= \ln\left(\frac{M^2}{\mu^2}\right)+\int_0^1 dx \ln\left(\frac{p^2}{M^2}x(x-1)+\frac{m^2}{M^2}x+(1-x)\right) \\
&\approx \ln\left(\frac{M^2}{\mu^2}\right)+\int_0^1 dx \ln\left(\frac{m^2}{M^2}x+(1-x)\right) \\
&\quad + \int_0^1 dx \frac{(p^2/M^2)x(x-1)}{(m^2/M^2)x+1-x} \\
&= \ln\left(\frac{M^2}{\mu^2}\right)-\frac{m^2}{M^2}\ln\left(\frac{m^2}{M^2}\right)-1-\frac{1}{2}\frac{p^2}{M^2}+O(M^{-4}).
\end{aligned}
\tag{11.33}
$$

[3]In general, after renormalization M will also depend on μ. However we have considered that $M(\mu) \approx$ constant close to the matching scale $\mu \sim M$.

Summing both contributions, $(1, a)$ and $(1, b)$, one obtains:

$$
\begin{aligned}
\mathcal{M}_R^{(1)} &= \mathcal{M}_R^{(1,a)} + \mathcal{M}_R^{(1,b)} \\
&= i \frac{\lambda^2(\mu)}{(4\pi)^2} \left[\frac{3}{2} \frac{m^2(\mu)}{M^2} \ln\left(\frac{m^2(\mu)}{\mu^2} \right) - \frac{1}{2} \frac{m^2(\mu)}{M^2} \right. \\
&\qquad\qquad\qquad \left. + 1 + \frac{1}{2} \frac{p^2}{M^2} - \ln\left(\frac{M^2}{\mu^2} \right) \right]
\end{aligned}
\tag{11.34}
$$

We will now move on to the computation of the diagrams for the effective theory. The transition amplitude of the first diagram reads:

$$
\begin{aligned}
\mathcal{M}^{(1,a)\text{eff}} &= \frac{1}{2} C_0^{(0)} \int \frac{d^D k}{(2\pi)^D} \frac{1}{k^2 - m^2} \\
&= i \frac{3}{2} \frac{\lambda^2}{(4\pi)^2} \frac{m^2}{M^2} \mu^{2\epsilon} \left[\frac{1}{\hat{\epsilon}} + \ln\left(\frac{m^2}{\mu^2} \right) - 1 \right].
\end{aligned}
\tag{11.35}
$$

The second one is simply:

$$
\mathcal{M}^{(1,b)\text{eff}} = -i C_2^{(1)} + i p^2 \bar{C}_0^{(1)}.
\tag{11.36}
$$

After renormalizing in the $\overline{MS}$ scheme, we obtain

$$
\begin{aligned}
\mathcal{M}_{\text{eff},R}^{(1)} &= \mathcal{M}_R^{(1,a)\text{eff}} + \mathcal{M}^{(1,b)\text{eff}} \\
&= i \frac{3}{2} \frac{\lambda_{\text{eff}}^2(\mu)}{(4\pi)^2} \frac{m_{\text{eff}}^2(\mu)}{M^2} \left[\ln\left(\frac{m_{\text{eff}}^2(\mu)}{\mu^2} \right) - 1 \right] \\
&\qquad\qquad\qquad - i C_2^{(1)}(\mu) + i p^2 \bar{C}_0^{(1)}(\mu)
\end{aligned}
\tag{11.37}
$$

where $\lambda_{\text{eff}}(\mu)$ and $m_{\text{eff}}(\mu)$ are the running coupling and mass in the effective theory (which, in general are not equal to the ones in the full theory). For the matching, in order to avoid potentially large logarithmic contributions from $\ln(M^2/\mu^2)$ we must choose the matching scale μ around M. Here we shall choose it $\mu = M$. The matching condition $\mathcal{M}_{\text{eff},R}^{(1)} = \mathcal{M}_R^{(1)}$ then gives:

$$
\begin{aligned}
\frac{3}{2} &\frac{\lambda_{\text{eff}}^2(M)}{(4\pi)^2} \frac{m_{\text{eff}}^2(M)}{M^2} \left[\ln\left(\frac{m_{\text{eff}}^2(M)}{M^2} \right) - 1 \right] \\
&\qquad\qquad - C_2^{(1)}(M) + p^2 \bar{C}_0^{(1)}(M) \\
&= \frac{\lambda^2(M)}{(4\pi)^2} \left[\frac{3}{2} \frac{m^2(M)}{M^2} \ln\left(\frac{m^2(M)}{M^2} \right) \right. \\
&\qquad\qquad\qquad \left. - \frac{1}{2} \frac{m^2(M)}{M^2} + 1 + \frac{1}{2} \frac{p^2}{M^2} \right]
\end{aligned}
\tag{11.38}
$$

Therefore we get to the following relations between the parameters of the full and effective theory

$$\lambda(M) = \lambda_{eff}(M), \qquad m(M) = m_{eff}(M), \qquad (11.39)$$

for the mass and coupling and

$$\bar{c}_0^{(1)}(M) = \frac{1}{2} \frac{\lambda^2(M)}{(4\pi)^2 M}, \qquad (11.40)$$

$$c_2^{(1)}(M) = -\frac{\lambda(M)}{(4\pi)^2} \left(1 + \frac{m^2(M)}{M^2}\right), \qquad (11.41)$$

for the coefficients. One must notice that the terms proportional to $\ln(m^2/M^2)$ cancel exactly in the matching. **This is a general feature of effective theories: both the complete theory and the effective one have the same IR behaviour (thus also the same IR divergences) and therefore, the IR divergences will always cancel in the matching.** In this case, if we wanted to take the limit $m \to 0$ for the light field ϕ, it would give rise to an IR divergent logarithm. However, this logarithm cancels in the matching and our effective theory is IR safe (well defined in the limit $m \to 0$).

Thus, after the matching we are left with the following effective Lagrangian

$$\mathcal{L}_{eff} = \frac{1}{2} \left[1 + \frac{1}{2} \frac{\lambda_{eff}^2(\mu)}{(4\pi)^2 M^2}\right] (\partial\phi)^2$$

$$- \frac{1}{2} \left[m_{eff}^2(\mu) - \frac{\lambda_{eff}^2(\mu)}{(4\pi)^2} \left(1 + \frac{m_{eff}^2(\mu)}{M^2}\right)\right] \phi^2$$

$$- \frac{1}{4!} \left[-3 \frac{\lambda_{eff}^2(\mu)}{M^2} + a \frac{\lambda_{eff}^4(\mu)}{(4\pi)^2 M^4}\right] \phi^4 + \dots, \qquad (11.42)$$

where a is determined by the matching given in the second line of Fig. 11.2 which has been left as an exercise. The running parameters $\lambda_{eff}(\mu)$ and $m_{eff}(\mu)$ can be calculated solving the RGEs with the initial conditions given by (11.39). The low scale is usually chosen as $\mu^2 \sim m^2$. Note that the Lagrangian (11.42) doesn't have a canonically normalized kinetic term. In order to obtain the canonical kinetic term and, therefore, use the standard Feynman rules we must perform a field redefinition:

$$\phi \to \phi \sqrt{1 - \frac{1}{2} \frac{\lambda_{eff}^2(\mu)}{(4\pi)^2 M^2}}. \qquad (11.43)$$

Thus, our final Lagrangian takes the form

$$
\begin{aligned}
\mathcal{L}_{eff} &= \frac{1}{2}(\partial\phi)^2 - \frac{1}{2}\left[m_{eff}^2(\mu) - \frac{\lambda_{eff}^2(\mu)}{(4\pi)^2}\left(1 + \frac{m_{eff}^2(\mu)}{M^2}\right) \right. \\
&\quad \left. - \frac{1}{2}\frac{\lambda_{eff}^2(\mu)m_{eff}^2(\mu)}{(4\pi)^2 M^2} \right]\phi^2 - \frac{1}{4!}\left[-3\frac{\lambda_{eff}^2(\mu)}{M^2} \right. \\
&\quad \left. + (3+a)\frac{\lambda_{eff}^4(\mu)}{(4\pi)^2 M^4} \right]\phi^4 + \cdots \\
&\equiv \frac{1}{2}(\partial\phi)^2 - \frac{1}{2}\bar{m}_{eff}^2(\mu)\,\phi^2 - \frac{1}{4!}\bar{\lambda}_{eff}(\mu)\,\phi^4 + \cdots
\end{aligned}
\tag{11.44}
$$

for which we can define the usual Feynman rules following the standard procedure and that we can use for calculating scattering processes like $\phi\phi \rightarrow \phi\phi$ with one-loop matching precision.

Further Reading

A. Pich, *Effective Field Theory*, arxiv:hep-ph/9806303

I. Stewart, *Effective Field Theories Lecture Notes*, http://pages.physics.cornell.edu/~ajd268/Notes/EffectiveFieldTheories.pdf

C. Scrucca, *Advanced Quantum Field Theory*, http://itp.epfl.ch/files/content/sites/itp/files/users/181759/public/aqft.pdf

A.J. Buras, *Weak Hamiltonian, CP Violation and Rare Decays*, arxiv:hep-ph/9806471

W. Skiba, *TASI Lectures on Effective Field Theory and Precision Electroweak Measurements*, arxiv:org/pdf/1006.2142v1.pdf

Chapter 12
Optical Theorem

Abstract The optical theorem can turn out to be a handy tool in calculating absorptive (imaginary) parts of self-energies (or scattering amplitudes) or it can be a useful cross check for some calculations. In this chapter we will explicitly deduce the standard form of the optical theorem and try to establish a robust and transparent notation. We will also take an illustrative example and recover the well known Breit-Wigner approximation for unstable particle propagators.

12.1 Optical Theorem Deduction

The scattering matrix $\mathcal{S}$ can be written as $\mathcal{S} = \mathcal{I} - i\mathcal{M}$, where $\mathcal{I}$ is the identity operator and $\mathcal{M}$ the transition matrix. The unitarity property of $\mathcal{S}$ guarantees that $\mathcal{S}^\dagger \mathcal{S} = \mathcal{I}$, to all orders in perturbation theory. Therefore, we can write the following:

$$\mathcal{S}^\dagger \mathcal{S} = (\mathcal{I} + i\mathcal{M}^\dagger)(\mathcal{I} - i\mathcal{M}) = \mathcal{I} - i\mathcal{M} + i\mathcal{M}^\dagger + \mathcal{M}^\dagger \mathcal{M} = \mathcal{I}. \qquad (12.1)$$

Thus:

$$i(\mathcal{M} - \mathcal{M}^\dagger) = \mathcal{M}^\dagger \mathcal{M}. \qquad (12.2)$$

Given an initial state $|i\rangle$ and a final state $|f\rangle$, we have

$$i\langle f|(\mathcal{M} - \mathcal{M}^\dagger)|i\rangle = \langle f|\mathcal{M}^\dagger \mathcal{M}|i\rangle. \qquad (12.3)$$

Introducing the closure relation in between $\mathcal{M}^\dagger$ and $\mathcal{M}$ on the RHS of (12.3) we get

$$i\langle f|(\mathcal{M} - \mathcal{M}^\dagger)|i\rangle = \widetilde{\sum_n}\langle f|\mathcal{M}^\dagger|n\rangle\langle n|\mathcal{M}|i\rangle, \qquad (12.4)$$

© Springer International Publishing Switzerland 2016
V. Ilisie, *Concepts in Quantum Field Theory*,
UNITEXT for Physics, DOI 10.1007/978-3-319-22966-9_12

where $|n\rangle$ is a complete basis of orthogonal states, and where $\widetilde{\sum}_n$ is defined as

$$\widetilde{\sum_n} \equiv \sum_n \frac{1}{(2\pi)^{3n_j}} \int \prod_{l=1}^{n_j} \frac{d^3 p_l}{2E_l}. \tag{12.5}$$

In the previous expression n_j represents the number of particles in the state $|n\rangle$. If we suppose that the initial and final states are the same, then the LHS of (12.4) reads:

$$i\langle i|(\mathcal{M} - \mathcal{M}^\dagger)|i\rangle = i\big[\langle i|\mathcal{M}|i\rangle - \langle i|\mathcal{M}^\dagger|i\rangle\big]$$
$$= i(2\pi)^4 \delta^{(4)}(\mathcal{P}_i - \mathcal{P}_i)\big[\mathcal{M}_{i\to i} - \mathcal{M}^\dagger_{i\to i}\big]$$
$$= -2\mathrm{Im}(\mathcal{M}_{i\to i})(2\pi)^4 \delta^{(4)}(\mathcal{P}_i - \mathcal{P}_i) \tag{12.6}$$

where we have introduced the following notation:

$$\langle f|\mathcal{M}|i\rangle = (2\pi)^4 \, \delta^{(4)}(\mathcal{P}_f - \mathcal{P}_i) \, \mathcal{M}_{i\to f}. \tag{12.7}$$

On the other hand, looking at the RHS of (12.4) we obtain the following:

$$\widetilde{\sum_n}\langle i|\mathcal{M}^\dagger|n\rangle\langle n|\mathcal{M}|i\rangle = \widetilde{\sum_n}\langle n|\mathcal{M}|i\rangle^\dagger \langle n|\mathcal{M}|i\rangle$$
$$= \widetilde{\sum_n}\Big((2\pi)^4\delta^{(4)}(\mathcal{P}_i - \mathcal{P}_n)\Big)^2 |\mathcal{M}_{i\to n}|^2. \tag{12.8}$$

Thus, combining (12.8) and (12.6) we obtain the standard form of the Optical Theorem:

$$-2\mathrm{Im}(\mathcal{M}_{i\to i}) = \sum_n \frac{1}{(2\pi)^{3n_j-4}} \int \prod_{l=1}^{n_j} \frac{d^3 p_l}{2E_l} \delta^{(4)}(\mathcal{P}_i - \mathcal{P}_n)|\mathcal{M}_{i\to n}|^2. \tag{12.9}$$

If the initial state is a two-particle state $(a + b)$, the theorem takes the form[1]:

$$-2\,\mathrm{Im}(\mathcal{M}_{i\to i}) = 2\lambda^{1/2}(s, m_a^2, m_b^2)\sum_n \sigma(a + b \to n)$$
$$= 2\lambda^{1/2}(s, m_a^2, m_b^2)\sigma(a + b \to all), \tag{12.10}$$

What the previous expression states is that, except for the $\lambda^{1/2}(s, m_a^2, m_b^2)$ factor and a global sign, the absorptive (imaginary) part of the transition matrix is equal (order by order in perturbation theory) to the sum of the cross sections describing all the possible final states originated by the initial two body scattering. Schematically:

---

[1] See Chap. 3 for definitions of decay rates and cross sections.

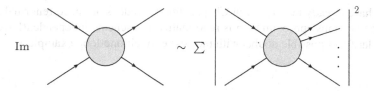

If the initial state is a one-particle state (a), then we can relate self-energies with decay widths:

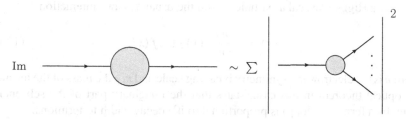

Therefore, the imaginary part of the self-energy is proportional to the total decay width of the particle:

$$-2\text{Im}(\mathcal{M}_{i\to i}) = 2\sqrt{k^2}\sum_n \Gamma(a \to n) = 2\sqrt{k^2}\,\Gamma(a \to all) \equiv 2\sqrt{k^2}\,\Gamma_a(k^2),$$

$$(12.11)$$

where k^2 and Γ_a are the squared four-momentum and the total decay width of the particle a. Note that in the previous equation we have written $\sqrt{k^2}$ instead of the mass m_a. This is because, on the LHS of the previous equation, we have the self-energy amplitude of a particle that is not necessarily on-shell ($k^2 \neq m_a^2$ in general). Thus, the total decay width must also be expressed as a function of $\sqrt{k^2}$ and not of m_a. This will be better understood with a latter example.

Therefore, (12.11) justifies the Breit-Wigner approximation of the propagator of an unstable particle. Consider for example the propagator of a scalar particle ϕ. After accounting for loop effects and renormalizing, the propagator contains the renormalized[2] two point function $\Pi_\phi^R(k^2)$:

$$\frac{i}{k^2 - M^2 + \Pi_\phi^R(k^2)} = \frac{i}{k^2 - M^2 + \text{Re}\,\Pi_\phi^R(k^2) + i\,\text{Im}\,\Pi_\phi^R(k^2)}. \qquad (12.12)$$

As a first order approximation, we can ignore the real part of the two-point function and we obtain the well-known expression:

$$\frac{i}{k^2 - M_\phi^2 + i\sqrt{k^2}\Gamma_\phi(k^2)} \approx \frac{i}{k^2 - M_\phi^2 + iM_\phi\Gamma_\phi}. \qquad (12.13)$$

---

[2]See Chap. 7 for more details on renormalization.

As we have already mentioned in Chap. 7, this term does not need renormalization thus, the previous approximation is renormalization scheme independent! We shall try to clarify all possible doubts with the next explicit one-loop example.

12.2 One-Loop Example

Consider a Higgs-like scalar particle ϕ with the usual Yukawa interaction

$$\mathcal{L}(x) = -\frac{m_f}{v}\phi(x)\bar{f}(x)f(x), \tag{12.14}$$

where v is the electroweak symmetry breaking scale and m_f the mass of the fermion. The optical theorem in this case, states that the imaginary part of the self-energy (given by a fermionic loop) is proportional to it's decay width to fermions:

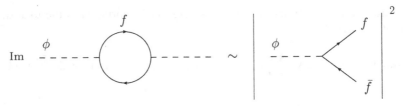

The decay width is straightforward to compute. It is given by:

$$\Gamma(\phi \to f\bar{f}) = \frac{1}{8\pi}\frac{m_f^2}{v^2}M_\phi\left(1 - \frac{4m_f^2}{M_\phi^2}\right)^{3/2}. \tag{12.15}$$

The self-energy is also rather trivial. Taking the following configuration of momenta

$$i\,\Pi(q^2) = \quad \begin{matrix} k+q \\ q \\ k \end{matrix} \quad \equiv -i\,\mathcal{M}$$

we obtain the following expression for the two-point function:

$$i\Pi(q^2) = -\frac{m^2}{v^2}\int\frac{d^Dk}{(2\pi)^D}\frac{Tr\{(\slashed{k}+\slashed{q}+m_f)(\slashed{k}+m_f)\}}{(k^2 - m_f^2 + i\varepsilon)[(k+q)^2 - m_f^2 + i\varepsilon]}. \tag{12.16}$$

Note that we have explicitly written down the $i\varepsilon$ regulator of the Feynman propagator because for this calculation it will play a important role (it should not be mistaken with $\epsilon = \frac{D-4}{2}$ where D is the number of space-time dimensions). Using the loop functions defined in the previous chapters we obtain

$$\Pi_\epsilon = \frac{12m_f^2}{(4\pi v)^2}\mu^{2\epsilon}\left(m_f^2 - \frac{q^2}{6}\right)\frac{1}{\hat{\epsilon}},$$

$$\Pi_R(q^2) = \frac{12m_f^2}{(4\pi v)^2}\left(\frac{q^2}{18} - \frac{m_f^2}{3} + \int_0^1 dx\, a^2 \ln\left(\frac{a^2}{\mu^2}\right)\right), \qquad (12.17)$$

with $a^2 \equiv -q^2 x(1-x) + m^2 - i\varepsilon$ and where we have separated the regular and the UV-divergent parts as in the $\overline{MS}$ scheme (for simplicity). We can now use this result to explicitly check on the optical theorem. The only piece of the self-energy $\Pi(q^2)$ that can develop an imaginary part is given bellow:

$$T(q^2) = \frac{12m_f^2}{(4\pi v)^2}\int_0^1 dx\, a^2 \ln\left(\frac{a^2}{\mu^2}\right)$$

$$= \frac{12m_f^2}{(4\pi v)^2}\int_0^1 dx\,[-q^2 x(1-x) + m_f^2]\ln\left(\frac{-q^2 x(1-x) + m_f^2 - i\varepsilon}{\mu^2}\right).$$

$$(12.18)$$

In order to find the imaginary part we have to find the roots of the equation:

$$-q^2 x(1-x) + m_f^2 = 0 \Rightarrow x_{1,2} = \frac{1}{2} \pm \frac{1}{2}\sqrt{1 - \frac{4m_f^2}{q^2}}. \qquad (12.19)$$

In the region in between x_1 and x_2 the only imaginary part of the logarithm is the one originated by the $-i\varepsilon$ term and it is $\pm i\pi$ depending on the sign conventions we adopt. Here we will choose it $-i\pi$. Thus

$$-\pi\frac{12m^2}{(4\pi v)^2}\int_{x_2}^{x_1} dx\,[-q^2 x(1-x) + m_f^2] = \frac{1}{8\pi}\frac{m_f^2}{v^2}q^2\left(1 - \frac{4m_f^2}{q^2}\right)^{3/2}. \qquad (12.20)$$

Comparing with (12.15), we finally obtain:

$$\mathrm{Im}\,\Pi(q^2) = \sqrt{q^2}\,\Gamma(q^2) = -\mathrm{Im}(\mathcal{M}), \qquad (12.21)$$

which is exactly what we were intending to prove.

Further Reading

C. Itzykson, J. Zuber, *Quantum Field Theory*

M.E. Peskin, D.V. Schroeder, *An Introduction to Quantum Field Theory*

V. Ilisie, *S.M. Higgs Decay and Production Channels*, http://ific.uv.es/lhcpheno/PhDthesis/master_vilisie.pdf

L.H. Ryder, *Quantum Field Theory* (Cambridge University Press, Cambridge, 1985)

T.P. Cheng, L.F. Li, *Gauge Theory of Elementary Particle Physics* (Oxford University Press, Oxford, 1984)

F. Mandl, G.P. Shaw, *Quantum Field Theory*

Appendix A
Master Integral

Consider the integral $J(D, \alpha, \beta, a^2)$ defined in (6.1):

$$J(D, \alpha, \beta, a^2) \equiv \int \frac{d^D k}{(2\pi)^D} \frac{(k^2)^\alpha}{(k^2 - a^2)^\beta}. \tag{A.1}$$

Given a four-vector k^μ we introduce the following Euclidean quantities (this is often called a Wick rotation):

$$k^\mu = (k^0, \mathbf{k}) = (i k_E^0, \mathbf{k}_E). \tag{A.2}$$

Therefore, we find the following:

$$k^2 = (k^0)^2 - \mathbf{k}^2 = -(k_E^0)^2 - \mathbf{k}_E^2 \equiv -k_E^2, \tag{A.3}$$

$$dk^0 = i dk_E^0, \tag{A.4}$$

$$d^D k = i d^D k_E. \tag{A.5}$$

We can thus, write (A.1) in terms of the previously introduced parameters

$$
\begin{aligned}
J(D, \alpha, \beta, a^2) &= i \int \frac{d^D k_E}{(2\pi)^D} \frac{(-1)^\alpha (k_E^2)^\alpha}{(-k_E^2 - a^2)^\beta} \\
&= (-1)^{\alpha - \beta} i \int \frac{d^D k_E}{(2\pi)^D} \frac{(k_E^2)^\alpha}{(k_E^2 + a^2)^\beta}.
\end{aligned} \tag{A.6}
$$

Let's now consider the following integral in spherical coordinates:

$$\int_{-\infty}^{+\infty} d^D x = \int d\Omega_D \int_0^\infty d|x| |x|^{D-1} \equiv S_D \int_0^\infty d|x| |x|^{D-1}, \tag{A.7}$$

© Springer International Publishing Switzerland 2016
V. Ilisie, *Concepts in Quantum Field Theory*,
UNITEXT for Physics, DOI 10.1007/978-3-319-22966-9

where we have defined $S_D \equiv \int d\Omega_D$ and where Ω_D is a solid angle. Using

$$\sqrt{\pi} = \int_{-\infty}^{+\infty} dx e^{-x^2}, \quad \int_0^{\infty} dx x^{D-1} e^{-x^2} = \frac{1}{2}\Gamma(D/2), \quad (A.8)$$

we obtain that $(\sqrt{\pi})^D$ can be written as

$$\begin{aligned}
(\sqrt{\pi})^D &= \left(\int_{-\infty}^{+\infty} dx e^{-x^2} \right)^D \\
&\equiv \int_{-\infty}^{+\infty} d^D x e^{-\mathbf{x}^2} \\
&= \int d\Omega_D \int_0^{\infty} dx x^{D-1} e^{-x^2} \\
&= \frac{S_D}{2} \Gamma(D/2).
\end{aligned} \quad (A.9)$$

Thus S_D is simply given by $S_D = 2\pi^{D/2}/\Gamma(D/2)$. Introducing the previous result we can now write (A.6) as:

$$\begin{aligned}
J(D, \alpha, \beta, a^2) &= \frac{2i(-1)^{\alpha-\beta}\pi^{D/2}}{(2\pi)^D \Gamma(D/2)} \int_0^{\infty} d|k_E| |k_E|^{D-1} \frac{|k_E|^{2\alpha}}{(|k_E^2| + a^2)^{\beta}} \\
&= \frac{2i(-1)^{\alpha-\beta}}{(4\pi)^{D/2} \Gamma(D/2)} \int_0^{\infty} d|k_E| \frac{|k_E|^{2\alpha+D-1}}{(|k_E^2| + a^2)^{\beta}}.
\end{aligned} \quad (A.10)$$

All we have left to do is to find the solution to the remaining integral

$$K \equiv \int_0^{\infty} d|k_E| \frac{|k_E|^{2\alpha+D-1}}{(|k_E^2| + a^2)^{\beta}}. \quad (A.11)$$

Performing a change of variable

$$z \equiv \frac{a^2}{(|k_E|^2 + a^2)}, \quad (A.12)$$

we find:

$$\lim_{|k_E|\to\infty} z = 0, \quad \lim_{|k_E|\to 0} z = 1, \quad dz = -\frac{2a^2|k_E|}{(|k_E|^2 + a^2)^2} d|k_E|. \quad (A.13)$$

Thus, we can write $|k_E|$ and $d|k_E|$ as

$$|k_E| = \left(\frac{a^2}{z}(1-z)\right)^{1/2}, \quad d|k_E| = -\frac{1}{2}dz(1-z)^{-1/2}(a^{-2})\left(\frac{a^2}{z}\right)^{3/2}.$$

$$\text{(A.14)}$$

The integral K in terms of the new variable takes the form

$$K = -\int_1^0 \frac{1}{2}dz(a^2)^{\alpha-\beta+D/2}z^{\beta-\alpha-D/2-1}(1-z)^{\alpha+D/2-1}$$

$$= \int_0^1 \frac{1}{2}dz(a^2)^{\alpha-\beta+D/2}z^{\beta-\alpha-D/2-1}(1-z)^{\alpha+D/2-1}. \quad \text{(A.15)}$$

Therefore the expression (A.10) becomes

$$J(D,\alpha,\beta,a^2) = \frac{i(-1)^{\alpha-\beta}(a^2)^{\alpha-\beta+D/2}}{(4\pi)^{D/2}\Gamma(D/2)}\int_0^1 dz\, z^{\beta-\alpha-D/2-1}(1-z)^{\alpha+D/2-1}.$$

$$\text{(A.16)}$$

Introducing the Euler Beta function,

$$\boxed{B(m,n) = \int_0^1 dz\, z^{m-1}(1-z)^{n-1} = \frac{\Gamma(m)\Gamma(n)}{\Gamma(n+m)}}, \quad \text{(A.17)}$$

taking $m = \beta - \alpha - D/2$, $n = \alpha + D/2$ we finally obtain the expression that we were looking for:

$$J(D,\alpha,\beta,a^2) = \frac{i}{(4\pi)^{D/2}}(a^2)^{D/2}(-a^2)^{\alpha-\beta}\frac{\Gamma(\beta-\alpha-D/2)\Gamma(\alpha+D/2)}{\Gamma(\beta)\Gamma(D/2)}.$$

$$\text{(A.18)}$$

Appendix B
Renormalization Group Equations

In the following we present a compendium of useful formulae regarding the renormalization group equations (RGEs) for the coupling constant, mass and a generic Green function.

B.1 Beta Function

Consider a model with one expansion parameter α, i.e., the QED Lagrangian introduced in Chap. 7. After including quantum corrections we need to perform renormalization. If α_0 is the bare parameter and α is the renormalized one, we have learnt that both parameters are related through a renormalization constant, say Z_α that we shall define as

$$\alpha_0 = Z_\alpha \alpha, \tag{B.1}$$

(in the QED case for example, we have $Z_\alpha = Z_3^{-1}$). As the renormalization constants are renormalization scale dependent (proportional to a $\mu^{2\epsilon}$ factor) and the bare parameter is μ-independent, we conclude that the renormalized parameter must also depend on μ so that the product $Z_\alpha \alpha$ is scale independent:

$$\boxed{\alpha_0 = Z_\alpha(\mu)\alpha(\mu)}. \tag{B.2}$$

In order to find the scale dependence of the renormalized coupling we differentiate with respect to μ on both sides of the previous equation:

$$\mu \frac{d\alpha_0}{d\mu} = 0 = \mu Z_\alpha \frac{d\alpha}{d\mu} + \mu \alpha \frac{dZ_\alpha}{d\mu}. \tag{B.3}$$

© Springer International Publishing Switzerland 2016
V. Ilisie, *Concepts in Quantum Field Theory*,
UNITEXT for Physics, DOI 10.1007/978-3-319-22966-9

The previous equation is called the RGE for the coupling constant. Defining the $\beta(\alpha)$ function as

$$\beta(\alpha) \equiv -\frac{\mu}{Z_\alpha}\frac{dZ_\alpha}{d\mu}, \tag{B.4}$$

we can write the RGE for α in the standard form

$$\boxed{\mu\frac{d\alpha}{d\mu} = \alpha\beta(\alpha)}. \tag{B.5}$$

The β function is called the **anomalous dimension of the coupling** and it can be perturbatively expanded as:

$$\beta(\alpha) = \beta_1\left(\frac{\alpha}{\pi}\right) + \beta_2\left(\frac{\alpha}{\pi}\right)^2 + \ldots = \sum_{i=1}\beta_i\left(\frac{\alpha}{\pi}\right)^i. \tag{B.6}$$

where the i index stands for the loop order. We can now proceed to find the perturbative solution to (B.5):

$$\begin{aligned}
\ln\left(\frac{\mu}{\mu_0}\right) &= \int_{\alpha(\mu_0)}^{\alpha(\mu)}\frac{1}{\beta(\alpha)}\frac{d\alpha}{\alpha}\\
&= \int_{\alpha(\mu_0)}^{\alpha(\mu)}\frac{1}{\beta_1\frac{\alpha}{\pi} + \beta_2\left(\frac{\alpha}{\pi}\right)^2 + \ldots}\left(\frac{d\alpha}{\alpha}\right)\\
&\approx \frac{1}{\alpha(\mu_0)}\frac{\alpha(\mu) - \alpha(\mu_0)}{\beta_1\frac{\alpha(\mu_0)}{\pi} + \beta_2\left(\frac{\alpha(\mu_0)}{\pi}\right)^2 + \ldots}
\end{aligned} \tag{B.7}$$

Rearranging terms we obtain the following

$$\alpha(\mu) = \alpha(\mu_0)\left[1 + \frac{\beta_1}{2}\left(\frac{\alpha(\mu_0)}{\pi}\right)\ln\left(\frac{\mu^2}{\mu_0^2}\right) + \frac{\beta_2}{2}\left(\frac{\alpha(\mu_0)}{\pi}\right)^2\ln\left(\frac{\mu^2}{\mu_0^2}\right) + \ldots\right] \tag{B.8}$$

Resumming the logarithms to all orders we get

$$\boxed{\alpha(\mu) = \frac{\alpha(\mu_0)}{1 - \frac{\beta_1}{2}\left(\frac{\alpha(\mu_0)}{\pi}\right)\ln\left(\frac{\mu^2}{\mu_0^2}\right) - \frac{\beta_2}{2}\left(\frac{\alpha(\mu_0)}{\pi}\right)^2\ln\left(\frac{\mu^2}{\mu_0^2}\right) - \ldots}}, \tag{B.9}$$

which describes the **running of the coupling** α **with the renormalization scale** μ.

B.2 Gamma Function

Let's now consider the mass renormalization. If the bare mass parameter is m_0 and the renormalized one is $m = m(\mu)$ we define Z_m as

$$\boxed{m_0 = Z_m(\mu)m(\mu)}. \tag{B.10}$$

Following the same procedure as before, we find the RGE equation for the renormalized mass to be

$$\boxed{\mu\frac{dm}{d\mu} = -m\gamma(\alpha)}, \tag{B.11}$$

where we have defined the gamma function (called **anomalous mass dimension**) as

$$\gamma(\alpha) \equiv \frac{\mu}{Z_m}\frac{dZ_m}{d\mu}. \tag{B.12}$$

Similar to the previous case it can be expanded as

$$\gamma(\alpha) = \gamma_1\left(\frac{\alpha}{\pi}\right) + \gamma_2\left(\frac{\alpha}{\pi}\right)^2 + \ldots = \sum_{i=1}\gamma_i\left(\frac{\alpha}{\pi}\right)^i. \tag{B.13}$$

Manipulating the expression (B.11) and inserting (B.5) we get to the following:

$$\frac{dm}{m} = -\frac{d\mu}{\mu}\gamma(\alpha) = -\frac{d\alpha}{\alpha}\frac{\gamma(\alpha)}{\beta(\alpha)}. \tag{B.14}$$

Integrating we obtain the following result:

$$\begin{aligned}
m(\mu) &= m(\mu_0)\exp\left\{-\int_{\alpha(\mu_0)}^{\alpha(\mu)}\frac{d\alpha}{\alpha}\frac{\gamma(\alpha)}{\beta(\alpha)}\right\} \\
&= m(\mu_0)\exp\left\{-\int_{\alpha(\mu_0)}^{\alpha(\mu)}\frac{d\alpha}{\alpha}\frac{\gamma_1\left(\frac{\alpha}{\pi}\right) + \gamma_2\left(\frac{\alpha}{\pi}\right)^2 + \ldots}{\beta_1\left(\frac{\alpha}{\pi}\right) + \beta_2\left(\frac{\alpha}{\pi}\right)^2 + \ldots}\right\}.
\end{aligned} \tag{B.15}$$

Truncating at the two-loop order we find:

$$\frac{\gamma_1\left(\frac{\alpha}{\pi}\right) + \gamma_2\left(\frac{\alpha}{\pi}\right)^2}{\beta_1\left(\frac{\alpha}{\pi}\right) + \beta_2\left(\frac{\alpha}{\pi}\right)^2} = \frac{\gamma_1 + \gamma_2\left(\frac{\alpha}{\pi}\right)}{\beta_1 + \beta_2\left(\frac{\alpha}{\pi}\right)} = \frac{\gamma_1}{\beta_1} + \left(\frac{\gamma_2}{\beta_1} - \frac{\beta_2\gamma_1}{\beta_1^2}\right)\left(\frac{\alpha}{\pi}\right) + O(\alpha^2).$$

$$\tag{B.16}$$

With the previous approximation we find

$$
\begin{aligned}
m(\mu) &= m(\mu_0) \exp \left\{ \int_{\alpha(\mu_0)}^{\alpha(\mu)} \left[\frac{d\alpha}{\alpha} \left(-\frac{\gamma_1}{\beta_1} \right) + d\alpha \left(\frac{\beta_2 \gamma_1 - \beta_1 \gamma_2}{\pi \beta_1^2} \right) \right] \right\} \\
&\approx m(\mu_0) \left[\frac{\alpha(\mu)}{\alpha(\mu_0)} \right]^{-\gamma_1/\beta_1} \left\{ 1 + \frac{\beta_2 \gamma_1 - \beta_1 \gamma_2}{\pi \beta_1^2} (\alpha(\mu) - \alpha(\mu_0)) \right\}, \quad \text{(B.17)}
\end{aligned}
$$

where we have used $e^{A+B} = e^A e^B$, $e^{B \ln A} = A^B$, and the approximation $e^\epsilon = 1 + \epsilon + O(\epsilon^2)$. Thus, the running mass with two-loop order precision is simply given by:

$$
m(\mu) = m(\mu_0) \left[\frac{\alpha(\mu)}{\alpha(\mu_0)} \right]^{-\gamma_1/\beta_1} \left\{ 1 + \frac{\beta_2}{\beta_1} \left(\frac{\gamma_1}{\beta_1} - \frac{\gamma_2}{\beta_2} \right) \frac{\alpha(\mu) - \alpha(\mu_0)}{\pi} \right\}. \quad \text{(B.18)}
$$

B.3 Generic Green Function

Finally let's consider an arbitrary Green function $\Gamma(p_i, \alpha, m)$, where p_i are the incoming/outgoing momenta. For simplicity we shall consider that the previous Green function only depends on one coupling α and one mass m. As usual, the bare Green function $\Gamma_0(p_i, \alpha, m)$ and the renormalized one $\Gamma(p_i, \alpha, m, \mu)$ can be related through a renormalization constant:

$$
\Gamma_0(p_i, \alpha, m) = Z_\Gamma(\mu) \Gamma(p_i, \alpha, m, \mu). \quad \text{(B.19)}
$$

The RGE for $\Gamma(p_i, \alpha, m, \mu)$ is then given by:

$$
\left(\mu \frac{d}{d\mu} + \gamma_\Gamma(\alpha) \right) \Gamma(p_i, \alpha, m, \mu) = 0, \quad \text{(B.20)}
$$

where we have introduced the generalized gamma function[1] as:

$$
\gamma_\Gamma(\alpha) \equiv \frac{\mu}{Z_\Gamma} \frac{dZ_\Gamma}{d\mu}. \quad \text{(B.21)}
$$

The dependence on the scale μ can be made more explicit by introducing the β and γ functions:

---

[1] It is worth mentioning that when using mass-independent renormalization schemes γ_Γ only depends on the coupling α and this is the case we are going to consider here.

$$\left(\mu\frac{\partial}{\partial\mu} + \beta(\alpha)\alpha\frac{\partial}{\partial\alpha} - \gamma(\alpha)m\frac{\partial}{\partial m} + \gamma_\Gamma(\alpha)\right)\Gamma(p_i,\alpha,m,\mu) = 0. \qquad \text{(B.22)}$$

In order to find the solution to this equation one can proceed as in the previous section and trade the μ dependence by α. We simply obtain

$$\Gamma(p_i,\alpha,m,\mu) = \Gamma(p_i,\alpha(\mu_0),m(\mu_0),\mu_0)\exp\left\{-\int_{\alpha(\mu_0)}^{\alpha(\mu)}\frac{d\alpha}{\alpha}\frac{\gamma_\Gamma(\alpha)}{\beta(\alpha)}\right\}, \qquad \text{(B.23)}$$

which can be perturbatively solved at any loop-order, just as in the previous cases.

Appendix C
Feynman Rules for Derivative Couplings

The formal steps to derive the Feynman rules from the interaction Lagrangian for *normal* vertices is straightforward and nicely presented in the literature. Here we shall only present a trick, useful to derive the Feynman rules corresponding to vertices that contain derivatives of the fields (and correctly assign the corresponding momenta). Consider the Scalar QED interaction Lagrangian given by

$$\mathcal{L} = ieA^{\mu}H^{+}\overleftrightarrow{\partial_{\mu}}H^{-} = ieA^{\mu}\left(H^{+}\partial_{\mu}H^{-} - (\partial_{\mu}H^{+})H^{-}\right). \qquad \text{(C.1)}$$

We know from Wick's theorem that H^{-} describes an ingoing particle or an outgoing antiparticle, also that H^{+} describes an ingoing antiparticle or an outgoing particle. Let's now consider we want to write down the Feynman rule corresponding to the following process $H^{-}(p) \rightarrow H^{-}(q)\gamma(p-q)$ shown in the Feynman diagram below

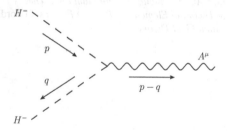

As we also know, in momentum space $e^{-ip\cdot x}$ describes an ingoing particle (antiparticle) with momentum p and $e^{ip\cdot x}$ describes an outgoing particle (antiparticle) with momentum p. Thus, substituting the fields from the interaction Lagrangian by their corresponding exponentials, assigning the momenta shown in the previous Feynman diagram we get

$$\mathcal{L} = iee^{i(p-q)\cdot x}\left(e^{iq\cdot x}\partial_{\mu}e^{-ip\cdot x} - (\partial_{\mu}e^{iq\cdot x})e^{-ip\cdot x}\right)$$
$$= e(p+q)_{\mu}. \qquad \text{(C.2)}$$

© Springer International Publishing Switzerland 2016
V. Ilisie, *Concepts in Quantum Field Theory*,
UNITEXT for Physics, DOI 10.1007/978-3-319-22966-9

Multiplying by an i factor (which comes from the S matrix expansion) we obtain the corresponding Feynman rule for this vertex

$$\boxed{ie(p+q)_\mu}.$$ (C.3)

Using the same trick one finds that $H^+(p) \rightarrow H^+(q)\gamma(p-q)$ is given by the following Feynman rule

$$\boxed{-ie(p+q)_\mu}.$$ (C.4)

This can be easily extended to any configuration of momenta and to any Lagrangian containing terms with derivative couplings. It very usual to find for example Feynman rules for triple gauge vertices (or similar) i.e., the triple gluon vertex given at the end of Chap. 5, where all momenta are incoming (which seems to violate momentum conservation). It is done like that for simplicity. Depending on the assignation of momenta for the process that one analyses, one must simply invert the necessary signs of the momenta to convert particles/antiparticles from incoming to outgoing.

Further Reading

L.H. Ryder, *Quantum Field Theory*, Cambridge University Press (1985)
A. Pich, *Effective Field Theory*, http://arxiv.org/abs/hep-ph/9806303
C. Itzykson, J. Zuber, *Quantum Field Theory*
A. Pich, *Quantum Chromodynamics*, http://arxiv.org/pdf/hep-ph/9505231v1.pdf
M.E. Peskin, D.V. Schroeder, *An Introduction to Quantum Field Theory*
T.P. Cheng, L.F. Li, *Gauge Theory of Elementary Particle Physics*, Oxford (1984)
F. Mandl, G.P. Shaw, *Quantum Field Theory*

Printed in the United States
By Bookmasters